HIDDEN SOLDIER

'He has led an extraordinary life. A great book – a truly incredible story'
Gerry Ryan, *2FM*

'A remarkable story of life and death in the world's most dangerous war zones'
Irish Independent

'It is a fascinating story and takes you to many parts of the world – from Bosnia to Iraq'
Matt Cooper, *Today FM*

'An incredible life spent in some of the world's most dangerous places – a great read'
Ireland AM

PÁDRAIG O'KEEFFE

Pádraig O'Keeffe is a native of Cobh, Co Cork, and a former Corporal in the French Foreign Legion. He served in the REG, the Legion's engineering regiment, and became a specialist in mine clearance and munitions disposal. His tours included UN and NATO duties in Cambodia and Sarajevo/Bosnia, as well as anti-terrorist patrols on the French-Spanish border. He was formally commended for his service in Sarajevo during the worst of the Yugoslav civil war. After leaving the Legion, he undertook five tours as a private security contractor in Iraq – eventually surviving one of the worst convoy ambushes ever mounted near Fallujah, in the Sunni Triangle. Badly wounded, Pádraig recuperated in Ireland before undertaking anti-kidnapping contracts in Haiti. He is currently negotiating further security contracts.

RALPH RIEGEL

Ralph Riegel is the southern correspondent for Independent Newspapers and covers the region for the *Irish Independent*, *Sunday Independent* and *Evening Herald*. He has also reported for the INM group from the US, UK, Europe and Australia. He is a regular contributor to RTE, TV3, BBC1 (NI), Channel 4, the *Independent* and the *Daily Telegraph*. He has previously worked with a variety of newspapers including *The Avondhu*, *The Nationalist*, the *Waterford News & Star* and the *Irish Examiner*. His first book was *Afraid of the Dark* and he also covered the Ian Bailey libel case for the updated edition of *Death in December*. Both books were bestsellers.

HIDDEN SOLDIER

An Irish Legionnaire's Wars

from Bosnia to Iraq

Pádraig O'Keeffe
with
Ralph Riegel

THE O'BRIEN PRESS
DUBLIN

First published 2007 by The O'Brien Press Ltd,
12 Terenure Road East, Rathgar, Dublin 6, Ireland.
Tel: +353 1 4923333; Fax: +353 1 4922777
E-mail: books@obrien.ie
Website: www.obrien.ie
Reprinted 2007.

ISBN: 978-1-84717-032-3

PHOTOGRAPHS
The photographs used in this book are either the author's own
or those of his friend Mick McCarthy.

British Library Cataloguing in Publication Data

O'Keefe, Padraig Hidden soldier :
an Irish legionnaire's wars from Bosnia to   Iraq
1. O'Keefe, Padraig
2. France. Armee. Legion etrangere
3. Soldiers of fortune - Ireland - Biography
4. Bodyguards - Ireland - Biography
I. Title II. Riegel, Ralph
355.3'54'092

2 3 4 5 6 7 8 9 10
07 08 09 10 11 12 13 14

Typesetting, editing, layout and design: The O'Brien Press Ltd
Front cover photograph (below): Getty Images;
other cover photographs are the author's own.
Printing: Creative Print and Design, Wales.

ACKNOWLEDGEMENTS

I want to thank my parents, Denis and Eileen, my sisters, Brid and Catriona, and my nieces, Caitlin, Erin, Saoirse and Juliette, who have borne the burden of my life in conflict. Their loyal support has been one of the core pillars of my life.

Thanks to my true friends here at home – you know who you are. I have opened my heart to you all and you have never faltered in your belief in me. You stood by me even when I unintentionally brought these conflicts home. There will always be a place in my heart for you all. To Keith Kenny and the long-suffering staff of the Rob Roy pub. To James Curry for having the courage to follow his convictions.

Special thanks to Tony O'Mahony and Paddy O'Keeffe who sat with me through long nights, which first ran into weeks and then months, slowly coaxing those painful memories from my darkest corners for this book. I love you like brothers.

To Mick McCarthy, who is a Legion brother, and has been through conflict at my side from Bosnia and Iraq to Haiti. A man whose life would truly be worth my own – a true hero.

To Kathy, Artemis and Alice of the Hart Group, who kept my best interests at heart. A special mention to the late Sean O'Donovan, who made the phone call that led to this book.

And thanks to Ralph Riegel for his patience and good humour over the past year.

Finally, a heartfelt thanks to the US Army Medical Corps, Cork University Hospital and South Infirmary Hospital – I owe you all my life. Special thanks to Dr Declan Pender and Dr Sean T O'Sullivan.

Pádraig O'Keeffe, Cork, July 2007

*　　*　　*

I am deeply grateful to Pádraig O'Keeffe for entrusting me with this book which became a remarkable journey for both of us. Working on this project was a great honour.

The book would not have been possible without the incredible and unquestioned support of my wife, Mary, my children, Rachel, Rebecca and Ralph, and my mother, Nora. None ever complained about having to watch me disappear, often for days on end, into books about the Foreign Legion, Iraq and Bosnia.

Similarly, my thanks to all within Independent Newspapers, in particular the various news desks, as well as numerous reporters and photographers in Cork, for their unstinting support. A special 'thank you' to Commandant Dan Harvey of the Southern Brigade for his honest and frank critique of the manuscript.

A special mention also to the late Sean O'Donovan for making the initial introductions which led to this project. Sean's death at such a young age was an absolute tragedy. This book would not have been possible without him.

A substantial '*merci*' – as always – to Michael O'Brien, Ide Ní Laoghaire, Emma Byrne, Ivan O'Brien, Claire McVeigh and all at O'Brien Press for making this book a reality.

Some surnames in this book have been shortened to their prefix and, in a handful of other cases, names have been deliberately changed. This is purely to protect the personal security of the individuals involved – for obvious reasons. But their stories have been unaltered and are faithfully recounted.

There are many fantastic books on the French Foreign Legion but, if you are to read only two, I can recommend Ewan McGorman's *Life in the French Foreign Legion* and Martin Wilmslow's *The Last Valley*. Both capture the essence, courage and mystery of *La Legion*. Robert Fisk's *The Great War for Civilisation* should be compulsory reading for anyone with even a passing interest in the Middle East.

In conclusion, a grateful mention to my late father, Ralph, ex-US Navy and USS Brooklyn – a quiet, unassuming man who first told me boyhood stories about the Foreign Legion and their famous Sidi-Bel-Abbès base in North Africa. Like so many of his generation, he served proudly in the company of heroes.

Ralph Riegel

The Earth is a very small stage in a vast cosmic arena. Think of the rivers of blood spilled by all those generals and emperors so that, in glory and triumph, they could become the momentary masters of a fraction of a dot. Think of the endless cruelties visited by the inhabitants of one corner of this pixel on the scarcely distinguishable inhabitants of some other corner. How frequent their misunderstandings, how eager they are to kill one another, how fervent their hatreds. Our posturings, our imagined self-importance, the delusion that we have some privileged position in the Universe are challenged by this point of pale light.
Our planet is a lonely speck in the great enveloping cosmic dark. In our obscurity, in all this vastness, there is no hint that help will come from elsewhere to save us from ourselves

Carl Sagan,
from a public lecture at Cornell University,
13 October 1993

This book is dedicated to the memories of Yves, Sean and all our Iraqi colleagues of Security Team Charlie 31 who died by my side on a lonely, dusty road outside al-Habaniyah on 7 June 2005; to Aki, Nick, Morne, Seb and Bill, who gave their lives so that others might live; to the Legionnaires and security contractors who made the ultimate sacrifice and continue to make those sacrifices in the most distant, hostile corners of the world to protect and defend the defenceless. It was an honour to have been amongst your ranks.

To Denis, who made it home. To the men and women of Hart Security and to the men of the French Foreign Legion.

My life has been blessed for I truly know what it is like to walk amongst heroes.

Pádraig O'Keeffe, Cork, July 2007

CONTENTS

1

THE MOST DANGEROUS PLACE ON EARTH

Our battered Toyota sped across the dusty road, keeping pace with the convoy vehicles in front that we were charged with protecting. That stifling noon of Monday, 7 June 2005 was already proving to be a nightmare. Our Iraqi drivers had been appalled that we were delivering a load through the 'Sunni Triangle' north of Baghdad, one of the most dangerous places on earth. Their fears and the condition of our ageing trucks had already slowed down the convoy as it went from our compound near the Abu Ghraib complex in Baghdad towards Fallujah, the heartland of the insurgency that was tearing Iraq apart.

Worse still, our route would, by necessity, take us north of Fallujah as we headed towards our final destination of al-Habaniyah, along a narrow, elevated road that had already proved itself one of the favourite hunting grounds for local insurgents.

As usual, our team leader, Yves M., had cried out '*Allah*

Ackbar' (God is Great) over the radio as we left the security compound that morning to try and encourage everyone, but the cry didn't get the usual hearty response and, when our Iraqi drivers answered half-heartedly, I knew that I wasn't alone in being desperately worried about this mission. But at least I had my ex-French Foreign Legion mate, Denis B., on the convoy security detail, along with Yves, an ex-French Army veteran, and Sean L., a former member of South Africa's private security Ronin outfit. Yves had also served in the Croation Army during the Balkans conflict, so he had combat experience and knew what he was about.

But no trip near Fallujah was ever without its risks, even for the American troops in their heavily armoured M1 Abrams battle tanks. The craters that littered the dusty roadside were a mute testament to the attacks that had already earned Fallujah its fearsome reputation. But today all we were escorting were trucks loaded with beds and billet equipment for soldiers of the new Iraqi Army, hardly the stuff that insurgents would target – if, that is, they knew what was inside the trucks. To them it could as well be ammunition or medical supplies. Or they might simply relish the chance to kill a few more infidels.

I sat in the rear of the speeding saloon and tried to maintain my focus. As my years of French Foreign Legion training had taught me, often it was the little things that kept you alive. I had two bags for my gear stowed carefully by my side, one for fighting from inside the car and one in case I had to get out of the vehicle in a hurry. Each rear door was draped with my body armour, ready to be grabbed at a moment's notice.

The heat was stifling and, as I gazed out the window, the whole countryside seemed to be burned brown. The only bit of colour came in flashes of green from a few stunted palm trees, and even they were hard to spot through the dust clouds. Fuck it, but there was dust everywhere. If you expected violence on arrival in Iraq, you certainly weren't disappointed, but no-one warned you about the dust, the flies and the heat.

On my lap was my AK-47 Kalashnikov assault rifle, with extra clips of ammunition stuffed into virtually every pocket and cubby-hole I could find. I was firm believer in the old infantryman's motto that you can never have too much ammunition. Directly in front of me was my Iraqi driver, Arkan H., and alongside him an Iraqi security contractor, Wisam D. They were both married men with families, and I knew they wouldn't be doing this job unless they really needed the money. But there wasn't much talking, we were all too busy staying focused on what was going on along the roadside.

Today, we had to protect ten old articulated trucks heading from our secure compound in Baghdad to an Iraqi police depot just outside al-Habaniyah. To deliver the trucks and their cargo we would have to cross the insurgents' own backyard. The key was getting there and back out again as fast and as discreetly as possible. The security detail involved five vehicles, mostly old, battered Opels, BMWs and Toyotas, like my saloon. These kinds of cars were far better for maintaining a low and discreet profile – if you drove a powerful new jeep you might as well have a neon sign over your head flashing 'foreign security contractor'. These civilian convoys were as hazardous as our job could get. And in a country gradually tearing itself apart with violence, this area between Abu Ghraib and Fallujah was the epicentre of Iraq's carnage.

I was assigned to the convoy's centre as the Counter Attack Vehicle (CAV). I had to ensure that the long line of trucks stayed together and, in the event of an attack, I also had to try and prevent any segment of the convoy being isolated and overrun. In an ambush, I was expected to respond to threats in any situation, and, if necessary, to give the insurgents something to think about while the rest of the convoy got out.

But just keeping the convoy together was proving difficult enough. The articulated trucks were all old, had been in constant use since the Coalition invasion and, most worrying of all, were prone to breaking down at precisely the wrong time. Not one of the trucks or security vehicles would have been allowed on a European or US road. And yet our very lives depended on these clapped-out artics.

Some 12km from al-Habaniyah, one of the trucks got stuck as we swung off the main supply road and on to an old dirt track. This was our worst nightmare. We were in bandit country, we had to deploy our Iraqi security detail – plus, we also had to cope with the possibility of the US military mistaking us for insurgents. Sure enough, as we desperately worked to get the truck moving, a US patrol swung on to the same dirt track and, spotting our armed Iraqi guards, instantly deployed in a defensive formation as if we were insurgents.

None of us underestimated the danger we now faced. US patrols had lost so many troops that some had naturally adopted a 'shoot first and ask questions later' policy when faced with possible threats. Before the American patrol could fully deploy, we had to order our Iraqi contractors back into the vehicles and Sean walked, hands held over his head, towards the lead US vehicle, an armoured Humvee, to try and explain our predicament. He walked slowly, his hands in clear view and repeatedly shouting, 'Friendly force, friendly force', not an easy task when you know a .50 calibre heavy machine-gun is trained directly on you. One blast from that weapon is enough to cut a man in half. But, after a few nervous minutes, the US patrol lowered their M16 assault rifles and, at last, acknowledged who we were.

I sat in the car and nervously fingered the trigger guard on my AK-47. The thought flashed across my mind that this was a strange place to find a chef from Cobh, Co Cork. Whatever about joining the Foreign Legion for a military life, Pádraig O'Keeffe certainly never thought he'd end up at thirty-five years of age sitting in a battered old Toyota with only a Russian assault rifle for protection in probably the most dangerous place in the world. I took a final nervous drag on my cigarette and waited for what was to come next.

It took us almost an hour to get the truck free of the soft sand and moving again, more than enough time to advertise our presence to any insurgents in the area. By the time we had all the trucks moving again, the traffic coming against us had become quite heavy. Within twenty minutes of our truck breaking down, I had noticed a steady increase in traffic on the road. We had never seen this level of traffic coming out of al-Habaniyah before – and, interestingly, the traffic was all moving against us, as if people were leaving the area for a specific reason.

With their previous ambushes and use of improvised explosive devices (IEDs), the insurgents had proved that they were evolving a very special skill in killing Coalition troops, as well as Iraqis working for the Interim Government and, most of all, the private security contractors, like myself, who undertook the security jobs no-one else wanted.

I smiled grimly as I realised that, far from being in the safest part of the convoy, the killing fields of Iraq had shown that the convoy centre was where the initial attack almost always came. The insurgents had previously favoured attacking a convoy at the front and rear, in the normal manner, thereby trapping it. But, because of how convoys in Iraq were being protected, the insurgents were now often attacking in the centre, hoping to cut the convoy in half.

Attacks were also mounted at the start and end of our missions as we left or returned to our Abu Ghraib compound: this danger was reflected in the fact the Iraqi police and foreign security contractors had erected a special security zone around the area, which required vehicles to pass through almost two hundred metres of three-metre-high blast walls which lined either side of the approach roadway, forming a kind of tunnel. This was supplemented by machine-gun posts and pill-boxes. And yet the attacks still came on a weekly basis and you had to maintain high alert even at the very end of a job.

Now, as we finally approached a hamlet just a few kilometres from al-Habaniyah, we drove along an elevated road with mud-brick houses on either side. Not much farther to travel, I thought, and we'd be okay. I kept a light grip on my rifle and fought the urge to re-check the magazine – I'd already checked it at least six times. Just one kilometre ahead we would have to make a tight left turn, enough to force us to slow our pace dangerously and spread out the convoy. As the turn loomed ahead, I repeatedly scanned the buildings for any sign of movement, because I realised that if I were going to hit this convoy, that is where I'd mount my attack.

The US patrol had already informed us that the track we were following towards al-Habaniyah was effectively no-man's land. One US Army Sergeant, in a quiet aside to our security team as we worked to free the stranded truck, advised us not to use the route and to head back to Abu Ghraib. He warned us that there were no Coalition forces in the area and that the hamlets around al-Habaniyah were believed to be rife with insurgents. While we were told that we would receive Coalition assistance in any emergency, none of our security team held out much hope of it in reality. We knew we were on our own. But what could we do? If we abandoned the trip and headed back to Baghdad, we'd only have to bring the convoy back out here again tomorrow. Better to take our chances and run with the convoy now.

As we approached the turn, out of the corner of my eye I caught sight of a figure huddled by the side of a mud-brick building to my left. I mentally registered his position and continued to scan the road ahead. I looked back to scan the house a second time and, sure enough, I could now clearly see the man crouching with an AK-47 in his hands. Instinctively, I knew that this was the moment we'd all dreaded. Time seemed to freeze as my Legion training began to take over.

The crouching figure slowly began to raise his arms and I could see the distinctive shape of the AK-47. I knew I had to signal the ambush to the rest of the convoy. We'd be cut to pieces if we ended up trapped amongst these buildings sitting on a elevated roadway some six feet above ground level.

Without thinking, I levelled my rifle and emptied a full magazine at the figure, seeing several rounds hit him. Almost instantly, I could hear a hail of fire erupt in reply from all around us. I shouted to Arkan to accelerate and get us away from the turn, and he was screaming back at me, 'Mujahedin, Mujahedin.' Seconds later, the windows of the Toyota exploded in a shower of fragments and I realised that Wisam was already dead, his bloodied head resting on the dashboard beside Arkan.

Suddenly I saw that the trucks ahead of me had stopped moving, and I instantly knew that something was seriously wrong. Our Toyota had barely rolled to a stop behind a truck when I ordered Arkan to sweep around it and move to the front of the convoy to find out what was wrong and why we were stopping in such an exposed position. Out of the corner of my eye, I spotted a civilian truck coming against us and, caught in the wrong place at the wrong time, it erupted in a hail of Kalashnikov fire from the insurgents. The truck ground to a halt and its cab partly shielded our saloon as we finally pulled up beside our own convoy artics. The poor civilian truck driver was dead within seconds. I didn't know it at the time but that civvy truck would probably save my life.

I hardly registered the sound of multiple 7.62mm rifle rounds crashing into the Toyota as I realised that, instead of coming from the left where I had shot one insurgent, the bulk of the ambush fire was now coming from my right. Within seconds of stopping, the Toyota was being carved apart by rifle fire and I knew that if I was to live through the next few minutes I had to get out of that car fast. Arkan was already out and taking shelter. As I kicked open the rear right door I felt a blinding pain in my elbow and realised I'd been shot. But with my injured arm I was still able to hold and fire the AK-47, and with my other arm I held my body armour in front of me as a shield. I could see the body armour flapping from the 'thud' of incoming rounds. I fell on to the roadway, still firing my rifle one-handed, and was instantly shredded by glass fragments that now littered the roadway.

I knew I had to find cover fast, so I combat-crawled on my elbows across the roadway, around the Toyota and into the shelter provided by my car and the stranded artics. I ignored the pain in my arms, only afterwards discovering that I'd torn lumps of flesh from my hands and arms as I crawled over the broken glass. But the incoming fire was incredible. I could hear 7.62mm rounds crashing through the car, the artics and even striking the roadway, tearing chunks of stone and clay up into our faces.

I shouted to Arkan to push the car underneath the first artic in the convoy for extra protection from the incoming fire. Incredibly, the young Iraqi managed to do it, pulling against the frame of the car door at the front and sliding the big Japanese saloon underneath the front bumper of the artic. The instant I heard the 'thud' of the Toyota wedging itself underneath the old truck, I opened fire again, emptying a full magazine at the distant buildings where the fire against us was most concentrated.

And then, from directly behind us, all hell broke loose ...

2

DREAMING OF BEING
A SOLDIER

I was born on 18 September 1970, the day the rebel guitar legend Jimi Hendrix died. I'm not sure if that had implications for my attitude towards authority in later life, but I like to think it may have been an omen!

Cobh is my home town – a town with strong military links. From a young age the whole idea of a soldier's lifestyle attracted me strongly, though neither my father, who was a member of An Garda Siochána, the Irish police force, nor my mother, who was a nurse in Cobh hospital, had military connections.

Over the years, it also dawned on me that I didn't really like the classroom. It's not that I wasn't smart – my early school reports from the local national school were all pretty good – the problem was that, like generations of youngsters before me, I liked having fun much more than doing my schoolwork. Then for two years I was sent as a boarder to Scoil na nÓg in Glanmire, about eight miles away. This was an Irish-speaking school and I loved every minute of my time there and got on really well. Whereas some kids struggle when they're away from their home environment and regard boarding school with a kind of dread, I seemed to thrive on it. That was another lesson, something else that attracted me.

Unfortunately, my time at secondary school wasn't quite so happy. I didn't help matters much by focusing more on enjoying myself than on doing what my teachers wanted. I wasn't a problem student, I just didn't like the classroom regime and, most of all, the homework that kept me away from my friends and sports. I was by now a fervent soccer fan and my favourite team was Manchester United. At times, my week revolved around 'Match of the Day' on Saturday night and the occasional live match. When it came to a choice between United and homework, there simply wasn't a contest! The teachers increasingly regarded me as a 'messer'.

My other great interest was the Vietnam War. I don't recall precisely what triggered my fascination with South-East Asia, but, by my mid-teens, I was tracking down every book I could to find about Vietnam and its famous battles, from Khe Sanh to the Mekong Delta and the TET offensive. Mention anything to do with Korea or the Vietnam conflicts and I was spellbound. Not only that, but I liked reading about the great military outfits of the world, from the US Marine Corps through to the Spetsnaz units of the former Soviet Union. By my mid-teens, I could describe virtually every military campaign from 1945 onwards.

Slowly, it began to dawn on me that this was the kind of life I would like to lead. I thought that being a soldier was a noble profession, and it was the kind of career that truly tested your mettle. I felt it asked hard questions of you – about your courage, your discipline, your loyalty and your commitment to a cause. Above all, I was fascinated by the whole idea of comradeship and the way men in military units formed a bond that went beyond race, creed or colour. Without realising it, I had made my career choice, even though the process of trying to fulfil it was still some years away.

I sat my Intermediate Certificate exams in 1985 but I couldn't wait for an excuse to quit school. By this stage, I'd got a weekend job working in the Commodore hotel in Cobh. My job was collecting empty glasses from tables in the bar and restaurant and loading them into the dishwasher. It was pretty mind-numbing stuff, but it seemed to me to pay reasonably well and, at sixteen years of age, it gave me a taste of what life outside the classroom could be like. I spent every waking moment trying to think up ways to get out of going to school. My parents had bluntly refused to tolerate me quitting without having either a job or an apprenticeship lined up. They made no secret of the fact they really wanted me to sit the Leaving Cert and work hard at getting into third level or a decent pensionable job like the Gardaí or civil service. I suppose it's what every parent wants for their child, but it simply wasn't for me.

One weekend I happened to hear that the Commodore was looking for kitchen staff and I asked the chef about my chances of getting a job. He put in a good word for me and the manager told me there was a job for me if I wanted it. It was like music to my ears, and I almost floated home to tell my parents about the wonderful, not-to-be-missed opportunity that had come my way! Reluctantly they surrendered, but only on condition that the job offered the prospect of a full-time career. So I quit school at last and went to the Commodore hotel as their youngest trainee chef.

In the mid-1980s Ireland was a pretty grim place to be. There weren't many jobs on offer, the economy was struggling and the rate of emigration threatened to match that of the late 1950s. A lot of guys with very good Leaving Certs and even college degrees were having to go overseas for work – and yet here was I with a full-time job and money into my hand each week! I thought I was the luckiest guy in town, particularly as Cobh was hit worse by the recession that most other places, with its old, heavy industries shutting down.

For me, the next year or so was like a dream. The work was hard, the hours were long and pretty unsocial, and, to be honest, the pay wasn't particularly great. But I was out of the hated classroom, I didn't have exams to worry about and, best of all, I had money in my pocket when I was heading out with my friends on my night off. Working in the town's biggest hotel also meant that I was known to the various barmen in town so I was looked after in terms of discounts and drink promotions.

Actually, I was surprised to find that I was quite good at cooking. I also took pride in what I was doing. I had come to admire the chef Gordon Ramsay, who demanded absolute passion of all his chefs: the very second I heard him use the word 'passion' about the dishes he was preparing, I was hooked. His approach was something I tried to apply in my own cooking and, as a result, I got on very well wherever I worked. In fact, passion is something I try to apply to everything I do in life.

By now I'd grown into quite a determined character. I wasn't afraid of hard work and I liked the honesty associated with it. With my background, I was always going to have a blue-collar outlook on life and, today, that's something I'm very proud of. I suppose you could say my approach to life in general began to solidify at this stage. If I worked hard, I expected exactly the same from others. I also respected people who spoke their mind even it was something I didn't particularly like to hear, and, over time, I grew to hate anyone who played politics and 'flip-flopped' with their opinions and loyalties. From pretty early on in life, I also realised that respect was something that had to be earned, never demanded. As for failings, well, there were plenty of those too. I could be stubborn and, like plenty of other young men, I thought I was pretty much infallible at times. I could also fall victim to the old Gaelic failings – drinking and talking too much. I even ignored good advice about not taking up smoking, and have pretty much paid the price for it ever since. But I was also protective of my family and I put great store in the trust of my friends. All in all, I was a typical Irish teen and revelling in my new-found place in the world.

I stayed for just over a year with the Commodore before I

got the chance to move to Cork city where one of the biggest hotels, then the Fitzpatrick Silversprings, was hiring staff. It was a higher profile position, the money was better, and it was an important advancement in my career as a chef. Best of all, I quickly became friendly with the manager, John Gately, and he began to help me along the career ladder. But what I didn't confide to anyone outside my family was the fact that, while I may have had the potential to become a good chef, I was by now only interested in a military career.

Sadly, my determination to escape from school hadn't made entry into the Irish Defence Forces any easier for me. Ireland was caught in the grip of a serious recession and the Government in power was determined to slash costs and public expenditure wherever possible. One of the things this hit was defence spending, and recruitment to the Defence Forces was only slowly, and very reluctantly, allowed. At the same time there was huge demand, given the security of employment that came with a Defence Forces job, and the army were able to impose strict academic criteria for their potential new candidates.

Needless to say, it seemed to help your case if you had a father, brother or uncle already in the forces. I had no-one directly connected with the Defence Forces – and a fascination with the Vietnam War wasn't likely to impress too many Irish army officers! Even so, with no Leaving Certificate and no contact within the Defence Forces, I went ahead and mailed my application. The interview, when it finally came, easily proved equal to my worst nightmare. The interviewing officer clearly wasn't impressed by what he was reading on the resumé in front of him. I answered his questions honestly and spoke about how I was committed to having a military career and felt I would be an addition to the Defence Forces. His answer still haunts me almost two decades later. He told me he didn't think I had what it took to be a soldier. He hurt me to the core of my being – but, in hindsight, I feel the man did me a favour as I wouldn't have wanted to serve under someone who could destroy a young man's hopes in that manner.

The only hope he held out to me was that, because I was working as a chef, I could perhaps work in the army as a cook. Some chance! A few weeks later, a letter from the Defence Forces arrived at my parents' house and, when my mother handed me the official envelope, I knew the contents before I even opened the letter. But, far from being devastated by this rejection, I felt angry and insulted. I knew I had what it took to be a soldier and, no matter what the demands, I'd prove it. I vowed to prove them all wrong.

I considered my options. There were three that I could think of: joining the US Marine Corps, the British Army or the Spanish Foreign Legion. Strangely, I never thought of the French Foreign Legion at this time. I had no interest in joining the British Army and felt that the Marine Corps was the most obvious and attractive choice. The Corps were almost always involved in overseas deployments and that's what I craved above all else. They were also one of the most impressively equipped forces, and any Marine could expect to be trained in the use of the most modern and high-tech military equipment. But, unfortunately, I quickly learned that the Corps, at that point, were not recruiting anyone who wasn't a US citizen – despite this I attempted to enlist when I was in Miami.

My next choice was the Spanish Foreign Legion, an outfit which ranks as tough as it is unknown. Traditionally, the Spanish Legion recruits from parts of the former Spanish empire, particularly from South and Central America. Its training regime is feared and its soldiers are reputed to be amongst the toughest in the world. When I was on holidays in Spain I decided to travel to a Legion recruiting office in Cadiz where I attempted to enlist, but they told me the Legion had been closed to foreigners for some years.

So it looked as if I was destined for a career in the kitchen. But the one positive thing about working as a chef is that there is never a shortage of work, and you can pretty much pick and choose the jobs to suit yourself. At that point, I liked being in Cork and being close to Cobh – and even though I was working in Cork city, I still lived and socialised in my home town where I had a great circle of friends.

One of the things about growing up in Cobh was that you had to learn how to defend yourself. Every port or harbour town has its own demands – the streets always seem a little tougher, a bit meaner. Cobh is no exception to that rule. For the most part, it's a quiet town, but, if you're involved in a row or a dispute, you had better be able to defend yourself, particularly if there are a few drinks involved in the equation. I've had more than my fair share of fights over the years, usually because of something silly being said or a few drinks too many being consumed. But, from my early teens, the one thing I realised about myself was that I was never intimidated by the size of a guy trying to 'sort me out' or even the number of guys standing behind him threatening me. I always stood my ground, and, in Cobh, I knew I had friends who would never stand by and watch me being taken on by a gang. But those were also innocent days when a fight meant a confrontation with fists and never with a bottle or a knife. And, invariably, in the following days the fight was forgotten about, grudges weren't carried and you usually found yourself mates with the lad again.

But I began to tire of life in the Silversprings kitchens. It was time to look around for a change. And, much as I loved Cobh, the thought of going somewhere else began to attract me. I remembered how happy I had been as a boarder at primary school and I figured that maybe getting out of Cork for a while would do me good. After all, I was just turning twenty years of age and thought that I was well and truly ready to see the world. The question, rather, was whether the world was ready for me? And, sure, if my move didn't work out, well, there was nothing stopping me from moving back home. And the blessing of being a chef was that there were always jobs on offer in some hotel or restaurant. So I decided to set a pattern for the rest of my life – I decided to take a chance.

3

EARNING THE *KEPI BLANC*

It's a long, hard road from a hotel kitchen in Cork to the French Foreign Legion recruitment base at Castelnaudary. But that was the journey I decided to take after a chance occurrence on a lazy summer afternoon.

After working at the Silversprings, I'd taken up a new position at the Slieve Russell Hotel and Country Club in Ballyconnell, north of Belturbet on the old Cavan to Enniskillen road. My old boss, John Gately, had transferred to Cavan and he had offered me the opportunity to go and join him. He assured me, before he left Cork, that it was a move I wouldn't regret.

There were a lot of reasons not to make the move – Cork was my home, and Cavan, well, it wasn't Dublin or London in terms of its nightlife. But I got on very well with John and he was a superb operator in the hotel business. I knew he'd be exceptional for the Slieve Russell, which already had a reputation as a good hotel. I knew, with John at the helm, I'd get on pretty well there and who knew where that might take me? I was itching to leave anyway. So I decided to take a chance and I swapped Cork for Cavan.

At first I wondered about the wisdom of the move. The hours were long, the pay wasn't great. But at least I was doing something I enjoyed and that was what counted for me. I also knew that the experience of being away from home was good for me, and Cavan, despite my preconceptions, was actually a lovely town with decent people and a thriving nightlife. For a twenty-year-old, it wasn't half bad at all.

Then, one day in early June 1991, I was assigned to a split shift. This effectively wrecked your entire day, though you did have a couple of hours off mid-afternoon to prepare yourself for the evening shift in the kitchen. I decided to go into Cavan to kill a few hours. Normally, I'd have gone for a few pints and read the sports pages to catch up on how my beloved Manchester United were doing. But, this time, I decided to just wander around the town. To this day, I don't know what possessed me to walk into a local bookshop; if I'm honest, I think it was fate guiding my footsteps. But, just seconds after I entered the shop, a book caught my eye. It was by Simon Murray and called *Legionnaire*, and as soon as I picked it up it transformed my life.

Flicking through the pages, I knew straight away that I had found my calling. It sounds ridiculous, but I was never so sure of anything in my life as the conviction in that instant that I wanted to be a Legionnaire. It was as simple as that. Up until that point, I didn't even know what a Kepi was.

But I was immediately fascinated by the Legion – its history, its legends and the fact that, at once, it opened up all the doors that I thought had been shut to me. While I laugh about it now, initially I assumed at first that to join up I had go to Sidi-Bel-Abbès, the legendary former home of the Legion in Algeria, which had been abandoned in the 1960s! But, checking quickly through the book, I realised I just had to get to Marseilles and, without thinking twice, I strode out of the bookshop and looked for the nearest travel agency. I was so excited I didn't even buy the book right then. Within five minutes of taking a short break from the hotel kitchen, I had booked a one-way ticket to Marseilles and the Legion depot at Aubagne. My life had just changed course. The main reason for the one-way ticket was to show myself how determined I was – though the fact that I was broke and wanted to save money on the return fare may also have been a factor!

I knew virtually nothing about the Foreign Legion except what I had picked up in the brief glance at the book: the fact that their famous cap was called a 'Kepi' and that the Legion had bases in Marseilles, Aubagne and formerly in Sidi-Bel-Abbès. The legendary discipline, the desertion rates, the physical demands of training and the gruelling overseas assignments – I knew nothing about those yet. Still, the discovery of the Legion and its Aubagne recruitment base was like Christmas come early.

I got back to the hotel at 5.00pm and immediately handed in my notice. I gave my boss the required seven days' notice and worked five of the next seven days. I was careful not to explain why I was going to France. I think I was worried that someone might laugh at me. I did manage to take one day off to get my gear ready for France – and a second day I reserved for going on the lash with my workmates.

I never once had second thoughts. It never even entered my head to consider whether I had made a mistake. I had no misgivings – it just felt so natural. It was a calling that I felt I had been waiting for all my life for. I know some people would take weeks or months to agonise over such a huge decision, but it took me less than five seconds to make up my mind and I've never once regretted it. To be honest, I was more than ready for it and it was no major decision to walk out of that bookstore and book a ticket for France and the Legion. I had just made the easiest decision of my life, yet I was about to step off the face of the planet as I knew it. I just couldn't wait.

A week later, a friend from the hotel drove me from Cavan to Dublin airport for the flight to Marseilles. I had told no-one what I was doing, especially not my parents, relatives or friends in Cobh. I suppose I was afraid that they'd try to talk me out of it. Even the friend who drove me to the airport, who worked alongside me in Cavan, only found out what I was doing en route as he wanted an explanation about why I was going to France.

When I left Dublin airport I had my life's belongings with me – a one-way airline ticket, a Cork City football jersey, shaving gear and the sum total of about IR£10 in cash. Not much for a twenty-year-old heading out on the greatest adventure of his life. On the flight, all I could think about was getting to the Legion base as fast as I could. I was very excited about the prospect of finally having a military career.

I had so many thoughts going through my head. How tough would it be? Would I be able to handle the physical and mental challenges? I had by now read the book on the Legion but, to be honest, I knew that the book only scratched the surface. Deep down I had a tremendous belief in myself. I knew I wanted to be a soldier, and I knew I had what it took. Once I stayed true to myself, I figured, I would be okay.

Marseilles is a tough town, and, surprisingly, the Legion is not particularly popular here. I didn't know it then, but wearing that white Kepi did come at a price. When I got off the flight, I didn't speak French, I had hardly any money, but I still hadn't a second thought about joining the Legion. I caught a bus from the airport into Marseilles and, after a few hours, managed to get the train to Aubagne, the main Legion recruiting base. Aubagne is about 17km east of Marseilles on the Toulon road and is home to the 1st Regiment (RE) of the Foreign Legion as well as being the Legion headquarters. As I got off the train at Aubagne, I had my first introduction to the Legion when I spotted two Legionnaires in full uniform returning from leave. I stood back and tried to soak it all in – I was overwhelmed and knew, in that very instant, that I definitely wanted to wear that Kepi. I promised myself that, in six months' time, I'd wear the same uniform and take my place in *La Legion*.

I wasn't being cocky. I guessed that basic training in the Legion would be tough. But it was tough in every elite military outfit, wasn't it? I promised myself that, no matter what I had to go through, I'd come out the other side a fully-fledged Legionnaire.

I made my way to the taxi-rank outside the station and simply said to the driver: 'Foreign Legion.' I didn't even

know the correct French title for the Legion but he knew
straight away what I wanted and drove me to First Regiment's base outside the town. This is known as 'The Mother
Regiment', and Aubagne is where the Legion motto, *Legio
Patria Nostra* (The Legion Is Our Fatherland), is inscribed on
the HQ wall.

As we approached the barracks, the driver, in very broken
English, explained that he would stop the car about a hundred metres from the gate. He explained that it would be far
better if I walked that distance to the barracks – apparently
recruits are not normally known to arrive by taxi. The base
was surrounded by a high stone wall – I couldn't see anything inside it beyond the guard post that marked the
entrance.

People join the Legion for many different reasons. Some
are running away from something in their lives – a crime or
a failed romance. Others are trying to prove something to
themselves – that they're tough enough or have conquered
their fears. But for me it was that I wanted to have a military
career and wear that white Kepi – as far as I was concerned I
was running towards something, not away from anything.

Walking through the gate, I wondered how many guys
had turned around at the sight of those gates and said: Fuck
this, it isn't for me. It never occurred to me to stop and have
second thoughts. That one hundred metres is one of the fast-
est walks I've ever made in my life.

The guard at the gate seemed to understand instantly
what I was about. After a few minutes, I was brought inside
the gate and, suddenly, from being surrounded by civilians I

was amongst Legionnaires, some wearing the white Kepi, others in green berets and combat fatigues. The officers, I noted, wore a black Kepi. The signing-up process took about two hours. I waited while a Sergeant was called and, when he finally arrived, I found myself being interrogated in pidgin English.

I had to produce whatever papers I had on me, including my Irish passport, and, a few minutes later, I was handed a threadbare Legion tracksuit and told to change out of my own clothes. I didn't know it at the time, but the moment I handed my passport to the Sergeant, I legally belonged to the Legion. Now, fifteen years later, you're officially Legion property when only you sign your five-year contract – but back in 1991 I belonged to them the minute my red passport disappeared into the buff Manila folder on the Sergeant's desk.

The very first thing you learned was that the Legion wouldn't tolerate bullshit. Seconds after I had handed over everything I owned, a second new recruit learned the hard way that the Legion demanded a specific code of behaviour from its members.

The guy had arrived, like me, to join up, but had left a couple of loose, disposable razors in his holdall. When the recruiting Sergeant put his hand inside the bag to examine its contents, he slashed his hand on the razors, and, in what seemed like a blur, his fist whipped out of the bag and delivered a haymaker to the side of the guy's face, knocking him out cold. I stood there and thought: Fuck! What have I let myself in for here?

I eventually spent three weeks in Aubagne, and it was very much an acclimatisation for Legion life. You learned to march, you learned to keep your gear in order, you learned to salute properly and you began your introduction to what it meant to be a Legionnaire. And I had my first introduction to the dreaded *corvée* – the cleaning routine. One Legionnaire I met years later reckoned that he spent more time wielding a mop and broom than he ever did a rifle. Aubagne taught me that the Legion takes cleanliness to almost psychotic levels. Being on *corvée* duties meant spending your whole day cleaning – first your barracks, then the floor, then the toilets and finally, perhaps, the entire parade ground. If one thing in Legion life was to be avoided at all costs, it was *corvée* duties, and, needless to say, that was impossible.

I found the training tough at the beginning. I had no military background – I had never thought of joining the Reserve Defence Forces at home. Now I was in basic training with guys drawn from virtually every army in the world. We had guys who were ex-Russian Spetsnaz special forces, former US Marines and former South African Army personnel. I reckoned some of these guys were as tough if not tougher than some of the training Sergeants. Yet here I was, a chef from Cobh, who had never touched, let alone fired, a rifle in my life. The most dangerous thing I'd ever handled was a boning knife as I filleted meat in the Slieve Russell kitchen. And now I was expected to be a Legionnaire on the same level as these guys.

The first lesson learned was also the most important in the Legion: keep your mouth shut and do precisely what you are told. And at the end of my second week in Aubagne I learned just how wise that policy was. In the bunk bed to the left of me in the barracks was a Romanian guy who really fancied himself as a hard man. For the previous few days, he'd been mouthing off about how easy the training was and how much tougher it had been for him in his previous army service. I kept my mouth shut, even though he was beginning to piss me and others off. One night, as we were all sleeping, three Caporals (Corporals) came into the billet, dragged this guy out of the bunk below me and proceeded to beat the shit out of him right in front of us. We were stunned – and the next day he was gone. We never saw him again.

As time went on, I began to make a few friends and learn the ropes of Legion life. One man – Corporal Jones from England – helped me out a great deal. 'Jonesie' marked my card: what to do, what to avoid and how to make the training that bit easier. He put me right on everything, and, in particular, on a Legion practice nicknamed 'Gestapo', which was aimed at ascertaining everything the corps needed to know about its new recruits. This was a series of interviews, which almost bordered on interrogation, about your previous life.

The first session lasted about three hours and was like an A to Z of your life to date. It was a grilling that went on for several further sessions over an entire week, and, if any aspect of your story didn't gel, you were out. The Legion interrogators were incredibly adept at identifying recruits who were trying to hide something in their past – in most cases a criminal conviction or an ongoing problem with the police. If they were suspicious, they could devote two entire sessions to asking the same question in about two hundred different ways. As Jonesie explained, one of the worst crimes in the Legion is to be caught lying, and it's not something that would be sorted out the easy way – at the receiving end of a fist or boot.

Traditionally, the Legion was willing to turn a blind eye to recruits who had misunderstandings with the law in their own countries – unless, of course, it involved murder. And even that qualification was a relatively recent addition. But the Legion insisted on *knowing why* recruits wanted to be Legionnaires and any criminal past was a crucial factor in that.

Luckily, I had nothing to hide and I wasn't running away from anything except, perhaps, the admission that I had worked as a chef and was afraid I'd be sent straight to the Legion kitchens. When it came to the second stage of the Gestapo interview, my answers were still the same and I think they realised I was genuine about why I wanted to join up. But I was two weeks at Aubagne when 'Jonesie' approached me to say my father had rung the Legion asking whether anyone by my name had tried to enlist. I simply told Jonesie that the Legion was my family now and I had no regrets or second thoughts about leaving life in Ireland. I was offered the chance to enlist under a different name but I said no, my name is Pádraig O'Keeffe, I have nothing to hide. And I'd keep my identity until it was confirmed as Legionnaire O'Keeffe.

Unfortunately, after I left Aubagne, I never again met

Jonesie. I always wanted to link up with him again, maybe buy him a few drinks because he was one of those guys who helped me a lot when I needed it most. Soldiers often say that the true strength of an army is found with its Corporals and Sergeants, and Jonesie was typical of that.

Ironically, the only issue I had with the Gestapo was that, after hearing about my rejection from the Irish army and my career as a cook, one Legion officer suggested, as I had feared all along, that I train as a cook for them. But this wasn't something I had travelled all the way to Aubagne for. I wanted a rifle in my hands not a chef's ladle.

After the Gestapo, we faced into a gruelling series of medical and fitness tests. The latter were tough enough to frighten even the most experienced recruits. Jonesie had warned me: keep running no matter how tired or exhausted you are – because if you collapse you have to start crawling around the parade ground. I also realised that the tests weren't as much about fitness as attitude – in the Legion, just when you think you're finished, you dig deep and find that extra reserve inside yourself. Not for nothing has the Legion unofficially adopted as one of it mottos: *Marche ou Crève* (march or die).

I was pretty fit arriving at Aubagne – but, my God, did I need it. Parades were followed by marches where we learned the unusual Legion marching style, which is done to a different rhythm and cadence that a normal army style. Then we usually had forced runs, all aimed at pushing us to the limit.

But the three weeks in Aubagne flew by. I passed all the tests with flying colours and, in fact, a lot of the time was spent simply hanging around waiting to see who was going to be thrown out. We all knew that the real training would start at Castelnaudary. But the Legion mystique had a heavy presence at Aubagne – the Legion museum was there alongside the monument to the Legion dead. You really got a feeling of belonging to something very special. In the years that followed, arriving back at Aubagne from leave was like coming home. For me, it was where I felt safe, where I had finally found my calling.

And, to be honest, I never thought of the training as raw brutality. We were involved in very serious business here – if any one person didn't do their job properly, others could die. So the training had to be realistic and tough. I think there was very much an ethos of: If you can't take what we dish out, then fuck off because you shouldn't be here. Another factor was that the Legion wanted teamwork and had very little time for show-offs. I remember at one point coming out into a corridor and seeing a former Spetsnaz soldier showing off by doing Bruce Lee-style kung fu kicks. But he too ended up on the receiving end of a few beatings, all aimed at bringing the high-flyers down a peg or two. There was the Legion way and the wrong way, and everyone had to learn that.

The funny thing is, surrounded by all these dangerous types, the humble Cork chef was more than holding his own. The Legion wasn't a dream for me any more, it was a daily reality. In fact, I think being so young and having no military experience actually helped me. I never thought I knew more than the training instructors, and I was so scared of making a show of myself that I focused hard on everything I was asked to do. I think I wasn't experienced enough to know just how intimidated I should have been. The Legion could mould me exactly as it saw fit.

Before moving to Castelnaudary, I got my first real token of Legion acceptance. Recruits who are processed through Aubagne and sent for basic training receive a red (*rouge*) tag for their epaulettes, a small sign that they've climbed the important first rung on the Legion ladder. When I got my *rouge*, it was fantastic – the only feeling greater than it was when I finally got to put on my white Kepi as a fully fledged Legionnaire.

But any sign of overconfidence was well and truly knocked out of us at Castelnaudary. Known to Legionnaires simply as 'Castel', the town is in the Aude *Department* and is located on the main Toulouse to Carcassonne/Narbonne road. Castel is home to the 4th Regiment of the Foreign Legion and is the main training base for the Legion. All eight thousand members of the Foreign Legion know that at Castel true Legionnaires are forged. Raw recruits are treated like raw iron – you are hammered, moulded and hammered again until you emerge from Castel like hardened steel. And, for that reason, it is feared. For the first time, we got to wear brand new combat fatigues complete with the green beret, though that initially lacked the full regimental headbadge. The trip by civilian train from Aubagne to Castel went by in a flash – we were all excited and a bit intimidated at what lay ahead of us. All the recruits were in a special carriage set aside for Legion use. It seems a bit ridiculous to admit it but, on the train, we were all like children heading to the seaside and wondering would we ever get there.

The training barracks is about 12km outside the town and, for the couple of hours while we travelled from Aubagne and then waited for the military buses to transport us out of Castel town to the barracks, our Sergeants were relaxed and chatty. It was a scorching hot July day and several Sergeants were enjoying a few cold beers. It actually crossed my mind that all the reports about the brutality of Castel were wrong. Hell, I thought, these guys are actually decent enough. Maybe the next four months won't be so bad after all.

But that mirage was well and truly shattered the instant we passed through the barracks guard-post later that after-noon. Our bus hadn't even slowed to a stop when the train-ing cadre stormed up to the coach, burst through the doors screaming like demons and dragged us all out on to the parade ground. There are times when I close my eyes now and can still hear the cursing and shouting – the complete mayhem – as the NCOs got stuck into us for the first time. We literally fell over each other in our headlong rush to get off the bus – and still we earned a few punches and kicks from the NCOs. All we wanted was to get out into the open as fast as we possibly could.

If Aubagne had given me a romantic image of the Legion, Castel taught me what being part of an elite military outfit was all about. That, and the price to be paid for it in training. For the first time, I wondered what the fuck I was doing here. It was mayhem – total and utter madness. There were maybe twenty different nationalities amongst us forty recruits, and, despite the daily language lessons, the majority of us still spoke very little French. And here we had some of the toughest NCOs screaming at us in French. We didn't know precisely what to do yet, but we all understood that we'd better find out fast or we were in deep shit for the next few weeks.

Castel pushed the boundaries for us recruits. We marched past the point of exhaustion, we sang the Legion marching songs until we were hoarse, we were dragged out of our billets while half-asleep for manoeuvres, and we began learning about our weapons until we could strip them down blindfolded. Even the smallest failure – misunderstanding a French word or failing to clean your kit properly – resulted in either a punch or a kick. Recruits had teeth knocked out, went on marches while black-and-blue from beatings and were usually cold, hungry and exhausted. But slowly we began to adjust to the Legion way of life. It was hell at the beginning when every single boundary was pushed to the limit. Each recruit discovered precisely what they could endure – and also what they could achieve.

We learned the Legion history: its famous battles, the legendary last stands and the honour that stood at the core of every regiment. The Legion was unique as a military outfit and intensely proud of the fact. For instance, no other army in the world would commemorate a battle where Legionnaires were not only defeated but virtually wiped out. However, because of the incredible courage shown against impossible odds at Camerone in Mexico on 30 April 1863, the battle is a huge part of Legion folklore. Just three

Legionnaires survived from sixty-two who refused to surrender to a Mexican army of more than six hundred. When
the three survivors were taken before the Mexican commander he was astonished, and stated that they were 'not
men but demons'. The French Emperor Napoleon III immediately decreed that Camerone Day be commemorated. The
Legion marks 30 April each year as Camerone Day (and so
do I still, to this day), and officers annually serve breakfast to
enlisted men as part of the Camerone tradition. The senior
officer killed that day, Capitaine Danjou, had only one hand
and his wooden prosthetic was recovered after the battle.
The hand is now revered at Legion headquarters in
Aubagne as one of the corps' most prized icons. But, despite
the history lessons at Castel, most of all what we ordinary
recruits learned was just how much punishment we could
take and just what wearing the white Kepi truly meant.

I was lucky in ways at Castel. There was a small group of
English-speaking recruits and while we were forbidden to
speak anything but French, we discreetly chatted amongst
ourselves about what was going on. I quickly became good
friends with an English guy by the name of Nick Peters.
Some time later I learned that 'Peters' was his Legion name
and his real name was Pears. He'd already served in the
British Army so he knew the military lifestyle pretty well.
He was also a decent bloke. We worked together pretty
well – and, though we didn't know it at the time, our paths
would cross again in Iraq in less happy circumstances. Of
the group of forty recruits, about sixteen were English-
speaking: myself, two other Irish lads, a few Brits and two

Danes who spoke fluent English. Strangely, we had only one American in our recruit cycle; he was a decent guy with the nickname 'Tungsten'.

We were assigned to a barrack room that held twelve recruits. Our training NCO, Caporal Santos, had his bunk in the same room. We quickly learned that Cpl Santos was a man obsessed – mostly with making our lives a total misery as he forged us into true Legionnaires. The man was an incredible soldier, but I also suspected that he might be a lunatic. Santos was French, and that spoke volumes about his level of commitment to the Legion. Whatever about foreigners joining the Legion, a French national who wants to wear the Kepi as an ordinary trooper pays a huge price for the honour. He has no option but to change his name, he is not allowed receive mail, he has no contact with his family, and, for the most part, the regular French Army attempts to treat him like shit. As if all that isn't enough, he generally has to enlist in the Legion under an assumed nationality, usually Belgian, Swiss or Spanish. Since the tradition had been for the Legion to take in anti-social and even criminal elements, it was seen as not being suitable for 'respectable' French recruits, and the attitude lasted.

Needless to say, it's different for officers. The top three or four officer cadets from France's famous military academy at St Cyr are offered a Foreign Legion posting, and most grab the chance. Nothing looks quite as impressive on a military resumé as having held a Legion command. The fact that Legion postings almost inevitably involve combat was an extra vital ingredient for officer promotion.

But for Cpl Santos and thousands of other Frenchmen, wearing the Kepi was all the proof that was needed of what they had achieved in life. Santos was on a promotion route within the Legion and that meant getting us recruits through basic training before he moved on to his course proper. Another Caporal attached to our recruit cycle was Australian – Cpl Hamilton-Byrne – and he was even more frightening than Santos. Tall, lean and one of the toughest bastards I ever met, Cpl Hamilton-Byrne pushed the training boundaries beyond anything we could believe. But we took it because we knew that he had taken precisely the same thing himself. Both Santos and Hamilton-Byrne were preparing for their Sergeant's courses. Our other NCO, Cpl Munch, was German and was also on a promotion cycle. Cpl Munch was much more approachable than the others, and, if you had a problem, he was the man to turn to.

And, by God, did we have problems. We learned to sing the French Foreign Legion marching songs, word-perfect, even though most of us didn't speak much French. One word wrong and, after a screaming session complete with a few punches and kicks to focus the mind, the entire recruit class would begin the song cycle again. The Legion even has its own book of marching songs: *Carnet du Chant*. Its marching songs are amongst the most famous of any army in the world and include 'Kepi Blanc', 'Schwarze Rose', 'En Algerie'. One of the very few pleasant things at Castel was watching and listening to a unit as they marched, all singing, across the parade ground. Maybe it was my imagination, but every back seemed to straighten and every eye seemed to clear as those marching cadences swirled around the barracks. Those moments made you feel proud to be part of *La Legion*.

But if singing was tough, learning French was even worse. Of course, some phrases you learned very quickly because the incentive was avoiding a punch or a kick. *'Bunune'* was a bad screw up, and you prayed that you weren't responsible. *'J'en ai pleins de couilles'* literally translates as having an overloaded pair of testicles, but in the Legion the phrase meant that a Caporal or Sergeant wants your guts for breakfast. The other key phrase that recruits learn is *'dégage'*, an admonition from a superior to 'fuck off' that is never ignored. *'Fait gaffe'* was a warning about dangerous behaviour and could range from not picking up cigarette butts properly on the parade ground to the incorrect unloading of live rounds from a rifle.

We were roused at 5.00am every day and, after a gruelling physical fitness regime, we began the basic infantryman's course. By the second week at Castel, I got my hands on my first rifle or, in Legion parlance, *'fusil'* – the FAMAS assault rifle. This had been the French Army's main weapon since 1975 and fired the NATO 5.56mm standard round. It weighed 3.6kg, lighter than either the Russian Kalashnikov or British Enfield assault rifles, and was the first 'bullpup' or shortened-length design adopted by any Western army. But I also had to learn the weapon's serial number. And, like a child, my rifle became my sole responsibility. One trace of dust or rust on that rifle and I paid for it with blood and bruises. I learned how to strip and clean that weapon blindfolded. More than sixteen years later, I can still reel off the serial number of that FAMAS. As far as a Legionnaire was concerned, the FAMAS was to be treated like an extension of your own body.

Punches and kicks weren't the only punishments. I once joked to a friend that the Legion's basic training was easy if you could only learn to shoot straight and iron well. The Legion were manic about uniform preparation and kit cleanliness, almost as if it was impossible to be a bad soldier if your uniform was spotlessly clean and ironed perfectly. We even had to measure with rulers the pleats on our shirts and trousers. It was an absolute nightmare. Once, when one of our NCOs was in a bad mood and drunkenly decided to take it out on us, I was forced to iron my kit five times in one night, with each failed kit inspection being followed by him dumping all our gear out the window. After ironing uniforms all night, a punch or a kick would almost have been greeted with relief.

And kicks were used frequently. One time I was sitting on my bunk, just inside the barracks door, reading my first letter from home, when Sgt Fuchineu walked in, his Kepi on his head. This automatically meant we had to jump to attention and salute, but I was so engrossed in reading my letter that I didn't notice him. The next thing I saw was the letter hurtling towards my face, propelled by the Sergeant's boot. As I lay on the bunk, dazed and bleeding, he calmly marched off. But I had learned a valuable lesson about always being aware of superiors in the vicinity.

We also learned to dread the 'moron' punishments such as

being forced to comb the entire parade ground for cigarette butts or being assigned to washing dishes or peeling potatoes. For some Legionnaires, particularly some guys from Eastern Europe, such duties were fine – they got paid precisely the same as the guys in the firing line, and they were in the Legion largely for the money. But for those of us who wanted to measure ourselves against the very best, such assignments were a dread. One of my biggest fears was being permanently assigned to barrack duties, and being kept away from the crack combat regiments.

One of the things I remember most about Castel was that the food was absolute shite – even when you were given time to eat it. We may have been in one of France's most famous regions for *gastronomie*, but the Legion reckoned we merited little more than swill (we could have done with Ramsey here!). The canteen was huge and us recruits had to queue for food while our NCOs were served first. We generally reached our table just as Cpl Santos was finishing his meal and, when he finished, we were ordered straight back out of the canteen. Most of us learned to eat quickly and without chewing – and to shove as much food as possible into our mouths, and even our pockets.

But if we thought Castel was bad, we were about to get the shock of our lives. In most armies, the so-called 'hell week' of survival courses and all-night marches takes place at the very end of your basic training. The Legion, as usual, did things their own way. At the end of our second week in Castel we were told we were being sent to 'The Farm' for three weeks. I think if we'd known precisely what was in store for us, none of us would have gone.

The Farm was a training complex in the mountains, about 30km from Castel. For almost a month, it became a living hell for all of us. No matter what else happened in my life, I knew nothing could ever be quite as bad as The Farm. At one point, we had four continuous days of duties with no sleep, only survival training, long marches and then arriving back to barracks to find NCOs crawling all over our kit and equipment, looking for faults.

One morning, a training officer woke us up at 5.00am by throwing a 'flash-bang', a type of smoke bomb, into our barracks room. We had only been in bed for an hour after completing a gruelling 72-hour endurance march through the mountains. There was dust residue from the explosion everywhere – in our hair, in our beds, even in our underwear. One hour later, we were told we were having a full kit inspection. When we failed – as everyone knew we would – all our gear, from our socks to our bedclothes, was thrown out a second-floor window into the slime and mud of the track below. It had been raining solid for twenty-four hours and the ground resembled a muddy farmyard. We were then told that, two hours later there was to be a fresh kit inspection, forcing us to collect and sort our gear, fight for space in the washrooms and then iron everything – only to fail the repeat inspection and have to go through the whole soul-destroying cycle again. It was total insanity. We hadn't a hope of making any of the kit-inspection deadlines. But that was precisely the point. We were being asked to do the impossible – and the training cadre wanted to see how we reacted to pressure, exhaustion, mayhem, futility and maybe even despair.

The Farm was all about your mentality, it wasn't about whether you could make it physically as a soldier. It was about discovering weaknesses and identifying those recruits who would invariably seek the easy way out. The Farm instilled a kind of attitude, a Legion outlook on how things should be done and why you never, ever quit. For some, The Farm became a kind of living nightmare and they would never allow themselves be put through a similar experience again. It was more than the bare concrete walls and ceilings – the whole experience became a measure of just how painful military life could be. It was a combination of physical exhaustion, mental fatigue, self-doubt and, for some, probably even a bit of fear. A few lads slammed doors on their fingers or deliberately tried to break bones just get out of The Farm. There is only one word of mine that can summarise the experience – hell.

In the Legion, you never gave up and if a Caporal or Sergeant thought you had that kind of staying mentality, you'd be okay. I think I proved that to them once in Castel when we went on a 30km march. We all had two pairs of boots – one for guard duty and one for everyday use. It was company exchange day when old equipment was swapped for new gear. I was on guard duty but I needed to change my boots and I asked a Legionnaire friend to bring me a new exchange pair. He obliged and left my new boots at the end of my bunk. I was on duty for twenty-four hours and then had to race back to my billet to get into combat fatigues ready for the route march. It was only then that I realised the boots my friend had got me were just a fraction too small.

By the 5km mark I had to start running around the marching column to keep my mind off the pain from my blistered and cut feet because it was the only way I could keep going. By 15km the pain was so bad I had to start hitting myself with the butt of my rifle to take my mind off my suffering legs. When it reached the point that I couldn't belt myself anymore, I asked my friends to slap me with their rifles just to keep my mind off the pain in the legs. At one point, I even threatened the Master-Caporal that if he didn't up the tempo of the march I'd kill him, such was the agony that I was in. The pain was so bad I think I passed beyond the borders of madness. But I kept marching and at times I ran because the pain was no different than it was at a slower pace. I was suffering beyond the point of endurance – and that was precisely the core reserve of a Legionnaire. When we finished the 30km march, my feet were like lumps of raw meat, and minutes after I took off my boots they swelled to double their normal size. But I knew I had another march to complete the next day. The following morning, I had to ask four Legionnaires to help me into the boots – three held me down while the fourth forced my boots back on my feet. By the end of the return march I was almost insane with the pain, and the Caporal in charge knew precisely what I had gone through. My commitment to the Legion ethos was never again in doubt. I had kept marching.

I realised that if I could survive The Farm I could survive almost anything. Those three weeks became a defining time in my life. I came back from The Farm a true Legionnaire. I had cut the mustard – I knew what was expected of me and I knew that I had what it took to make not just a good soldier but a good Legionnaire. I realised with quiet pride that the Kepi would fit Pádraig O'Keeffe like a glove.

A few weeks later, there was a short ceremony at Castel where we were finally given our white Kepis. It still ranks as one of the proudest days of my life. To some people it is probably just a hat, but for those of us who sweated blood and tears in training and endured hunger, beatings and exhaustion, it was a symbol of passage. We weren't recruits any more, we were Legionnaires. There was no longer any distinction between us and the older Legionnaires – all we lacked now was the experience, rank and a regimental assignment. But there were constant reminders not to get cocky, and more than a few of us who came back from The Farm got a few slaps and punches for not saluting properly, for being caught speaking English or for no other reason than the Caporal or Sergeant thought we needed it.

Nor did wearing a white Kepi mean any change in the regime, particularly from the bedrock of any army, the Sergeants. At this stage we trained under a pretty elite group: Sgt Saito who was Japanese, Sgt Kazmarski who was Polish, Sgt Engler who was German and Sgt Fuchineau who was French. In the two decades after World War II, the Legion had boasted large contingents of German soldiers who had a huge influence; even some of the Legion marching songs were German. But in some ways Kazmarski, being a Pole, seemed to represent the future of the Legion which was drawing more and more recruits from Eastern Europe. Saito was like a military machine – he was the kind of sergeant you would follow anywhere in battle. And Fuchineau believed that the English and Irish made the best and worst Legionnaires: we were incredible in the field and in action, but caused absolute havoc when confined to barracks. For that reason, he took a special interest in Irish and English recruits, and, if I owe my Legion mentality and valour to anyone, I owe it to Fuchineau. But I paid the bill in pain and suffering.

And then, finally, our four months of basic training at Castel were over. In the final weeks, our regimental assignments were sorted out. Like all the other recruits, I went to Castel fancying myself as a 'para'. If the Legion was a legend, well then, the two parachute regiments, both of which were based in Corsica, were almost mythological in their own right. The paras had proven their mettle at innumerable battles, including Dien Bien Phu, and were, we thought, the elite within the elite.

Only the first five recruits in each class got to pick their regimental assignment, though the Legion did try to accommodate assignment requests where possible. As we neared the end of our training cycle, several Legion officers approached us and asked us to consider the Engineering Regiment which was developing several special outfits. I had finished fifth in my recruit class and was one of those approached. I was initially sceptical – I wanted to be a soldier and not an engineer. Shit, if I wanted that I could have gone to University College Cork or Cork Institute of Technology back home.

But the officers were enthusiastic and they pointed out that, contrary to belief, it was the Engineering Regiment that was first on any Legion combat roster. And the type of engineering they had in mind had more to do with demolition than construction. The 6th REG (Regiment Étrangère Genie) provided special explosives teams, anti-mine units, amphibious attack squads and even a special helicopter assault unit. Recruits would be trained in virtually every known type of explosive, and the overseas assignments were particularly rich pickings for the REG.

I chatted it over with Nick Peters/Pears, who had also made the top five of the recruit class, and we finally decided to give it a try. And so I opted for the 6th REG whose base was at Avignon in the South of France. I couldn't wait to get there, as much to get on with the real life of a Legionnaire as to get away from Castel and its training regime. But the Legion is nothing if not relentless – and, the night before I was to go on regimental assignment, one of the NCOs drunkenly staggered into our barracks at 10.00pm and decided to do an impromptu kit inspection! The end result was that we were washing and ironing our uniforms until 5.00am, and then had to catch a military transport at 7.00am to Avignon.

And so it was on to Avignon. After once providing a refuge for the popes from the violence of medieval Rome and its warring factions, Avignon now had to contend with hormone-filled young Legionnaires returning from combat duty in Africa, South-East Asia and the Middle East. The Legion base was about 20km from Avignon, and, in a town with a large Arab population, Legionnaires were often about as popular as the Ebola virus.

The Legion's tension with the Arab population derived from almost a hundred years of colonial wars in North Africa. The final straw was the loss of Algeria in 1961 and the Legion's famous home base at Sidi-Bel-Abbès. I quickly learned that there was mutual antipathy between Legionnaires and Arab men, and vicious fights were common in Avignon and other southern French towns where the two were forced to live cheek-by-jowl. Ironically, it was very similar in Corsica where the Legion parachute regiments were based. Corsicans didn't like Legionnaires and, if you were off duty, it was never wise to head into town drinking on your own.

We had less than three weeks of regimental introductory courses in Avignon before we were sent to northern France where the 6th REG were on manoeuvres with the rest of the Legion combat teams and elements of the French Army and Marines. This was my introduction to the Legion's never-ending-cycle: training manoeuvres, combat duties, back to base and then a short period of leave before it all started over again.

I also quickly learned that my initial preconceptions about the engineers and sappers couldn't have been more wrong. We were trained in all aspects of high-explosives: how to position charges, how to defuse mines, how to judge a fuse so that you didn't blow yourself up, how to trigger a defensive perimeter with anti-personnel mines – and all the while learning the ropes as combat infantrymen as well. The REG insisted its members be fully trained in amphibious warfare and also be able to support helicopter assault teams.

It may sound idiotic but, to a twenty-one-year-old fresh into one of the world's elite fighting forces, there is nothing quite like the thrill of getting to blow shit up in training. In Ireland, the most expensive piece of equipment I worked with was a €500 cooker – here I was entrusted with military hardware that cost millions. Our helicopter assault teams used Aerospatiale Puma choppers that cost almost € 20million each. The Legion's Cavalry units trained alongside the LeClerc main battle tank that equipped the French Army – and cost almost €10 million each. Even our FAMAS assault rifles cost thousands of Francs.

There was a serious edge to our training. We realised that if we didn't do our jobs properly a lot of people would die – starting with ourselves! So the concentration on explosive types and fuse timers in training was pretty intense. The Legion had a reputation for having the best 'sappers' in the business, and we were determined to prove it true.

Over time, Legionnaires would also go on manoeuvres to Mont Louis, one of the toughest commando training courses in the world. If you passed your training regime here, you could wear your badges with pride. Mont Louis, located high in the mountains, was used by the Legion to fine-tune the military skills we had started learning at The Farm and then at regimental level in the various Legion bases. As a commando training centre, Mont Louis was unsurpassed. It pushed men to limits they never thought they'd reach. But it also fostered the ideas of team-work and operating as a tight-knit military unit. If The Farm was a blunt object, Mont Louis was the precision instrument which sculpted soldiers into Legion specialists. Mont Louis honed the skills that the Legion so prized – that nothing was impossible to a well-motivated Legionnaire and his unit. We scaled cliffs that a person would normally be nervous simply to look at. We climbed obstacles and barricades using nothing but muscle, sweat and blood. We crawled through fields of barbed wire and undertook forced marches with full packs. We hurtled down the 'death-slide', a rope absail system that truly lived up to its nickname. And when we left Mont Louis, we walked that little bit taller and felt the Kepi sat that little bit prouder on our heads.

Yet, despite the danger and the strict training regime, life in the Legion was all that I had dreamed about and more. The discipline was still strict and the regimental sergeants and Caporals were ruthless in enforcing the training and operational codes. If you screwed up, you knew there was a price to pay – and often your entire outfit would pay the price as well. So you learned to hold your end up, and, in the days and weeks that followed my posting to Avignon, I made great friends and comrades.

I made a few enemies too! Some of our regimental sergeants were the toughest in the entire Legion. One man – Sgt Manges – was renowned for the sadistic pleasure he took in making a Legionnaire's life a misery. Unfortunately, I made the mistake of getting on Sgt Manges's wrong side. I was ordered to join the REG who were on manoeuvres in the North of France and I had just arrived there to report to the

Company Captain. As I stood outside his office in full uniform waiting to be called, a guy in a tracksuit walked by. I
didn't know who he was, I had never seen him before, he
wore no insignia and I didn't even know if he was a Legionnaire. So I didn't salute – and, without warning, the figure in
the tracksuit stopped, turned and hit me a punch straight in
the face. He then walked away without saying a word. After
picking myself up from the dirt, I then had to walk into the
Captain's office and make my first report with blood pouring from my face. My introduction to Sgt Manges was something I wouldn't forget in a hurry.

But, on the positive side too, we were Legionnaires now
and not raw recruits. That meant we began to have a social
life – and going for a few beers was not only allowed, it was
almost expected by our Caporals. Often in the bleak Legion
barracks, going for a few Stellas or Kronenbourgs was all
there was to do.

For almost six months I'd known nothing but barracks life.
And then, suddenly, we were given leave. Admittedly it
was only to a Legion-operated club in Marseilles but, as far
as we were concerned, it could have been Stringfellow's
Niteclub in London. We had five precious days – and they
passed in a drunken haze despite the fact we had less than
€20 each per day to exist on. We felt that if ever a group
deserved to let their hair down – and we all had shaven
heads by this stage – it was us. There was always a tight
bond within the Legion; it was like an unruly, occasionally
violent family. If recruits or ordinary Legionnaires didn't
have enough money for a few beers, invariably their

Caporal would look after them. It was a favour Legionnaires were expected to return when they were finally promoted.

And then we got sensational news. Less than two months after getting out of Castel and completing our combat manoeuvres in northern France we were off on an overseas assignment – to Cambodia, where the Legion were on UN election-support duties. I didn't know it at the time, but the Cambodian assignment was the first UN duty ever assigned to the Legion.

I had finally arrived. I knew guys who were two years in the Legion and were still waiting for an overseas posting, and here I was, just six months after I walked into the Legion base at Aubagne, and I was heading off to South-East Asia wearing my white Kepi! I wasn't just a soldier. I was a Foreign Legionnaire. And I was heading to South-East Asia as part of the first French military mission to the region since Dien Bien Phu.

4

ON DUTY IN CAMBODIA

It felt as if we were flying to Cambodia via the North Pole. The flight seemed to last forever, and we also had our first introduction to the fact that while the Foreign Legion was undoubtedly France's most famous military unit, we were definitely bottom of the pecking order when it came to resources from Paris.

My regiment, the 6ᵗʰ REG, was travelling from Avignon to Sihanoukville in Cambodia on an old Aeroflot aircraft specially chartered for the Legion. Needless to say, they weren't particularly designed for comfort, and we all knew they probably came cheap. The planes were so old they had to refuel several times on the long flight. We ended up stopping en route at Cyprus, Saudi Arabia, India and Bangkok before finally landing at Sihanoukville on the Cambodian coast facing the Gulf of Thailand. We arrived itching to limber up, and, for those of us on our first overseas assignment, hoping for a little action.

The Legion arrived in Cambodia in January 1993 as part of France's commitment to a United Nations military mission aimed at ensuring democratic elections went ahead without interruption or intimidation. It was very much a case of *déjà vu* for the Legion because Cambodia, like Vietnam and Laos, were once a critical part of the Legion's area of operations. This was indeed 'Indo-Chine' – and the Republic's involvement in the area went back so far that French was almost as widely spoken as English, which had become common because of the US involvement. Indo-Chine was still so much a part of Legion lore that one of our marching songs, '*Contre Les Viets*', was all about fighting the Vietminh back in the 1950s. Perhaps the greatest evidence of French influence in the region was that Cambodia's most lethal armed group used a French name: the Khmer Rouge.

Now, we were the new Legion and none of our personnel could trace their service back to a time when tours were conducted in Cambodia, Vietnam or Laos. But Legion history is never forgotten, and, just a few hours from our mission assignment area, across the Vietnamese border and north towards the Red River and China, lay a legendary battlefield: Dien Bien Phu.

If a military unit is defined by its history, then the Legion's shadow is dominated by Dien Bien Phu. The cream of the Legion's colonial parachute regiments had fought a savage fifty-seven-day siege against Vietminh forces led by General Vo Nguyen Giap before finally surrendering on 7 May 1954. The battle was the start of the death knell for France's colonial empire. Once again, the Legion had suffered a major defeat – but again the raw, defiant courage shown during the battle shone like a beacon. When Dien Bien Phu had its air-strip closed by the heavy artillery fire of the surrounding Vietminh forces, almost to a man the Legion paras, as well as non-paratrooper-trained Legion personnel not at Dien Bien Phu, volunteered to be dropped in as reinforcements despite knowing they almost certainly faced death or capture. Many who volunteered were not even jump-qualified as paratroopers. It was the kind of legacy every modern Legionnaire was expected to live up to. Part of the price of keeping the *Kepi Blanc* so special.

We were initially based at Sihanoukville, a coastal city named after the Cambodian royal family but which also went by the name Kompong Song. I was assigned to the Regiment's 1st Amphibie-Recon Company, so I knew I was going to see a lot of river patrols. The airport was right beside the coast and the famous beaches of the Gulf of Thailand stretched in front of us. But this wasn't a holiday, as our Caporals and Sergeants were quick to remind us. Despite the heat and humidity, we underwent a punishing daily fitness regime, all aimed at getting us ready for the hard work when small Legion teams would be sent out across Cambodia in support of election workers. For the next few months the countryside – the 'boonies' – would be our home.

I grew up reading books on the Vietnam War and this was an assignment I would almost have paid someone to let me be involved in. There is something intoxicating about the Far East – the smells, the heat, the colour, the jungle and the incredible warmth of the local people. The Legion generally has a policy of not fraternising with locals, largely because it can affect your ability to do your job, not to mention the fact that it can also compromise security.

But I couldn't help but like the Cambodians. Despite all the horrific things they had suffered, they were still friendly, warm and welcoming people. Almost everywhere we went there were the stark, brutal reminders of the country's recent past: temples where piles of human skulls were still stored; urban areas still struggling to recover from the deprivations of Pol Pot and his Khmer forces who had tried to brutalise the country back into the Stone Age. The last thing these poor people needed was foreign troops acting like assholes around their country. There isn't a Cambodian family which wasn't affected by the horrors of almost twenty years of warfare and, if the Legion could help ensure fair elections and offer them some hope for the future, I was proud to play my part.

But Legionnaires will still be Legionnaires. We weren't twelve hours on the ground when our unit had located a suitable local bar that offered cheap local beer. We knew we had to enjoy an opportunity while it lasted because our briefings had confirmed that, in a few days, we'd move out into small-unit assignments divided between the coast, the Mekong Delta, the Elephant Mountains and the Cardamom Highlands.

For the next few months I felt more like a construction worker than a soldier. My REG was assigned to reconstruction projects, and that meant Legionnaires swapping combat fatigues for work clothes as we repaired roads, rebuilt bridges, helped dam paddy fields and generally did whatever strategic 'Good Samaritan' work the local communities needed. The Khmer Rouge, under Pol Pot, had virtually dismantled the entire national infrastructure even to the point of destroying all wooden bridges throughout the countryside. Typically, the destruction had no strategic impact whatsoever on the incessant fighting between the Khmer, their local opponents and the Vietnamese. The only people who suffered were the ordinary Cambodian farmers. So the Legion decided to repair the bridges and reconnect villages that had been cut off from each other for almost a decade. But while it may have sounded noble and generous in theory, it was to prove back-breaking in reality.

The heat and humidity of Cambodia made any outdoor work extremely strenuous. Usually I was with a four- or five-man construction team working on bridges and culverts. Most Cambodian bridges were made of heavy timber frames and each had to be lifted into place using nothing more than sweat and brawn. We had no access to heavy construction equipment which would have made such jobs pretty straightforward. So we did it the old fashioned, muscle-and-sinew way. Beams had to be manually hauled off flat-bed lorries specially requisitioned by the Legion. These huge hulks of timber then had to be cut and shaped for a precise fit. Then they were individually levered into place by the construction teams. The conditions were such that we often worked in nothing more than boots and shorts – everything was totally drenched in sweat from the humidity. In many cases, Legionnaires spent the day up to their waists in water trying to secure the beams from underneath.

The most advanced tool we had available to us was a

petrol-driven chainsaw, which was supplemented by shovels, hammers, pick-axes and the odd bow-saw. The work was long, hard and tedious, yet we slowly but surely got the bridges repaired and re-opened. You could see the difference such work made to the lives of ordinary people. A simple task like moving from one village to another had suddenly become much easier. Farmers hauling fodder or rice didn't have to negotiate huge caverns in the road or ford a river where a bridge once stood. People were grateful for the improvement, and the smiles on the villagers' faces spoke volumes about what we were doing. The work may have been back-breaking but it was also inspiring. If ever anyone deserved a little good fortune and outside help, it was the Cambodians.

But the construction work didn't fool anyone about the true capability of the Legion. As the weeks wore on, our reconstruction tasks began to dwindle and, with the election looming, we began to 'fly the flag' with a planned series of nationwide tours and patrols. This was what we had come to Cambodia for, and, after more than two months, most of us were relieved to be able to swap the hard work of building bridges for the duties of a soldier. Initially, our patrols were either motorised or via patrol boat. Three-man teams would take to the roads on Toyota 4x4 pick-ups, with a 7.62mm belt-fed machine-gun mounted on the rear, and speed around the countryside. We knew we were only a token force, but we had to show the colours and try and persuade local groups that the elections would go ahead as planned.

For someone in his early twenties, it was adrenalin-pumping stuff. We also knew that, during the election itself, we would be responsible for the armed escort of ballot boxes to central counting stations. I particularly loved the amphibious missions – sweeping up small rivers and through the delta to villages that you'd swear had never before been reached. It reminded me of something straight out of *Apocalypse Now* or *Platoon*. During such patrols, a number of REG personnel would work closely with Legion infantry regiments. On such assignments I was responsible for handling the amphibious patrol craft, but I also had to be able to hold my own during a 10km or even 15km foot patrol with the infantry in dense jungle, which made movement a nightmare. In Legion parlance it was *'pied, pied, pied'* – 'foot, foot, foot'.

The most serious risk we initially faced was that of old, forgotten minefields. Cambodia had basically been a war zone for more than forty years and, like Angola, ranked as one of the worst countries for landmine injuries. The savagery of the fighting had resulted in hundreds of thousands of mines being laid, often at indiscriminate locations close to villages, schools or markets. One of the first things you notice in the Cambodian countryside is the large number of amputees in most villages.

On one of my first patrols I got graphic evidence of just how dangerous an unwary footfall could be. We were patrolling an area of jungle when we had to cross a small stream-bed. As we marched down, our Sergeant called 'Halt' and, sure enough, half-exposed in the mud of the riverbed lay an old landmine. The rainfall in Cambodia is so heavy during the monsoon season that mines are regularly exposed when the mud covering them is washed away. That's when an unsuspecting villager or child walks along and their life changes irrevocably. Our policy was either to safely defuse and remove the mines or destroy them in place.

Often, we would be asked to try and help the victims of landmines. Once I was back at our compound when a local villager stepped on a landmine and suffered serious damage to his leg and arm. His friends carried him, bleeding and screaming, to our compound so our doctor could treat his wounds. I happened to be passing by the open door of the surgery as the villager was being treated and, spotting me, one of the sergeants called me to help. I stepped reluctantly into the room where the poor guy was being treated, only to be handed a shattered finger. 'Get rid of that – we can't reattach it,' the medic grunted at me. I stared at the bloodied piece of flesh and thought: What the fuck do they expect me to do with this? I quickly withdrew from the room and decided straight away I'd throw it in the shitter, a deep pit on the outskirts of our compound perimeter that doubled as a garbage dump and latrine. I tried to scrape some dirt over the finger and then I gratefully went back to the compound, silently commiserating with the unfortunate villager.

However, about an hour later I heard a roar: *'Merde —* O'KEEFFE.' I ran towards the sound of my Sergeant's voice only to stop, dumbfounded, as I saw him pointing at the unit's mascot, a mongrel dog, who was standing in front of the barracks wagging his tail, the amputated finger which he had extracted from the disposal pit in his mouth. To make matters worse, the dog had obviously climbed out of the refuse pit via the latrine section because he was now covered in human excrement. In grim resignation I followed the direction of the Sergeant's finger and realised that I now had to recover the finger from the shit-covered dog. Then, when I buried the finger again, the dog circled around and I knew he would simply dig it up, so I threw it into the paddy fields – but that led to the same shout: '*Merde*, O'KEEFFE.' Finally, I threw the finger into an old minefield behind the barracks – and that dealt with both the finger and the hungry dog. But it was the last time I ever walked near the compound surgery when anyone, villager or Legionnaire, was being treated.

Undoubtedly the most legendary Legion NCO with us in Cambodia was Master-Sergeant Gercer. He was almost a twenty-year veteran and had been part of the famous Kolwezi operation in the Congo in May 1978. Gercer, and other paratroopers from the Legion's 2nd REP (Regiment Étrangère Parachutiste), had successfully subdued rebels who had taken over the area and held hostage a number of Europeans working in the mining industry. The rebels had been backed by Angola and had taken over the entire region of Kolwezi. Zaire, worried about regional stability, had asked for French help. The operation was a tremendous success, with the Legion overcoming incredible odds and defeating the rebels. Despite that, six Legion paras, 250 rebels and 170 European hostages died.

Master Sgt Gercer was as tough as nails, totally unforgiving of mistakes and one of the meanest bastards I have ever met in my life. Having said that, he was the type of Legion-naire you always wanted to be. To some, he was the ultimate Foreign Legion type. If you were ever in a fire-fight, he was the guy you wanted leading your unit. Unfortunately, he seemed to have an abiding hatred of all English-speaking Legion recruits. He made no secret of this dislike and it made no difference that I was Irish and not English. We were all the same as far as the moustached veteran was concerned. I knew that Sgt Gercer was a Yugoslav, but I never found out precisely why he hated English-speakers.

Unfortunately, I broke my own rule about not attracting attention to myself with Master Sgt Gercer, and for the rest of my time in his area of operations he took an evil interest in everything I did. During one of our first rural postings in Cambodia we had established a fortified camp north of Krong Koh Kong, miles from anywhere. One night I was on sentry duty, manning a squad machine-gun, when I heard a noise in the jungle. I issued the required challenge, got no reply, heard the noise again and, when a second challenge went unanswered, I thought to myself: Fuck this – I'm not taking any chance. So I opened up with the machine-gun. Within seconds the whole camp was in uproar. Our Captain sprinted up to me in his underpants carrying a 9mm pistol and ready for action. 'O'Keeffe, O'Keeffe! *Merde*! What are you shooting at?' he screamed.

Unfortunately, it wasn't long before everyone realised it was a false alarm and Master Sgt Gercer gave me an evil look that meant I was in the deepest of deep shit. A few wits commented that I thought I was a machine-gunner on an

American Huey gunship and openly reminded me that the Vietnam War had ended twenty years before. They were relieved because they knew that the tough Yugoslav now had a target for all his attentions over the next weeks – and it wasn't them! For months afterwards, Master Sgt Gercer applied almost Castel standards to everything I did, from my kit to my patrols, from my parade-ground uniform to how I maintained my rifle.

When we rotated back to Sihanoukville, the duty officer issued my FAMAS rifle to another Legionnaire for use on a routine night patrol. Lazily, he returned the weapon to the armoury without wiping it down, a silly mistake in a country like Cambodia where the humidity took a terrible toll on all metal objects. Two days later, the weapons were being handed out by serial number and, when my FAMAS was handed out, tell-tale rust spatters were already visible on the barrel where it had not been cleaned. Master Sgt Gercer froze, stared at the weapon and, just as I stepped forward to take it, he caught the weapon by the barrel and, wham, swung it butt first into my face. I staggered and tried desperately to stay on my feet as the blood sprayed everywhere. Then I took the weapon and turned away. Not a word was spoken. Not only did my head feel like it was going to explode but I had to clean my rifle and then wash and iron my combat fatigues which were spattered with blood. A few hours later, I got that Legionnaire in a quiet corner and cursed him to his face. He kept trying to apologise to me but all I could hiss back was: 'You lazy bastard, all you had to do was wipe it down.' But that was the Legion – for every action there was a consequence. The mistake cost him, though – the price was a case of beer. That was how we settled our debts.

Yet men like Gercer were what gave the Legion its aura. His kind had helped make the Kepi what it stood for. A few years later, after I had earned my Caporal's stripes, I was walking across the parade ground at Avignon when I spotted Gercer, now promoted to Adjutant, striding towards me. My uniform was spotless, I was proud of my recent promotion and I flashed out a stiff salute to him. In the Legion, a salute to an NCO or officer that you have served with on an overseas tour – and whom you think deserves it – is accompanied by a brief greeting: *'Respect, mon Adjudant.'* After saluting and paying my respects to Master Sgt Gercer he saluted back and offered the merest hint of a smile. But to me it was worth pure gold.

Some may have regarded UN election support duties as an easy assignment, but the reality on the ground was anything but that. The tension within Cambodia was almost palpable. All around us were armed groups – Cambodian Army, Khmer Rouge and even drug gangs. And all of them had been blooded in one of history's longest and most vicious conflicts. Human life had little or no value here. Often, a team of four or five Legionnaires on patrol could be confronted by armed groups of twenty or fifty men. We could be 40km from our nearest Legion colleagues, and that would usually be a team of only four or five Legionnaires like ourselves. But we were the Legion, we wore our shoulder patches with pride and we never let anyone think we were intimidated. Still, deep down, we knew that if the Khmer or even the Cambodian Army wanted to take us out, we hadn't a snowball's chance because of our small units and the isolation of teams spread over huge areas.

Most of our difficulties occurred with the Cambodian Army and the police, both of whom had lucrative drug and smuggling interests on the side. They appeared less concerned about our involvement with protecting the democratic process than with any potential threat we might pose to their money-spinning off-duty incomes. And drugs were everywhere. Every rural village seemed to be engaged in harvesting 'weed' – it was as common out there as grass silage is back in Ireland. It was little wonder, I thought, that the US Army had such huge drug-addiction problems with its personnel when it finally withdrew its troops from Vietnam in the early 1970s.

Once, on a routine foot patrol in the middle of nowhere, we suddenly came out of the jungle and into one of the biggest fields of marijuana I have ever seen. It stretched for miles, almost as far as the eye could see. It was a junkie's paradise. But within seconds of our five-man combat team coming out into the open, we were confronted by about thirty Cambodian Army personnel. They were visibly pissed-off and immediately levelled their weapons at us. We responded, and, as the Mexican stand-off continued, tensions mounted, with us shouting for the Cambodians to lower their rifles and the Cambodians screaming back at us. We weren't going to back down and, even though there were only five of us, we made it perfectly clear that if there was going to be shooting, we were going to take a lot more than five of them with us.

Eventually, it dawned on the Cambodian troopers that we were merely on a security patrol and that the huge field of hash, which was clearly theirs, was of absolutely no interest to us. Each Legion patrol had an assigned interpreter and our interpreter worked overtime that day to ease the tensions and re-assure the nervous Cambodian troopers that we wouldn't interfere with their hash. We eventually walked warily away from each other.

But it highlighted the danger of our mission – if gunfire had erupted we were in serious trouble as we were outnumbered, outgunned and miles from any hope of support. In fact, if we had been killed it was quite likely that no-one would ever have even found us or discovered what had happened. When you're in the boonies in Cambodia, it is every man for himself.

Unfortunately, I also learned first hand about the true risks of the boonies when I fell seriously ill. But for the quick thinking of a Legion medic I could have died, so serious was the malaria infection I'd contracted. We were miles north of Sihanoukville when, one morning, I knew something was wrong. My head hurt, every muscle in my body seemed to be aching and, when I tried to shrug it off, the pain just kept getting worse. Finally, I told my Sergeant that I needed to check with the medic and, minutes later, when the medic took my temperature, he almost freaked out. My temperature was soaring over 103 degrees and the medic immediately told me I'd got malaria and needed to get to a hospital fast. He got on the radio and, within an hour, I was being loaded on to a Canadian Huey chopper for urgent transfer to a military hospital in Sihanoukville.

As sick as I was, I can still remember the thrill of that ride south in the legendary Vietnam-era Bell helicopter. The side doors were fully open and the Cambodian countryside of paddy fields, jungles and river deltas flashed by below. If I wasn't so ill I'd probably have asked the pilot to circle around for a while to let me enjoy the spin. But I was almost unable to speak and my temperature was now threatening to rage out of control. The medical team waiting for us at the hospital loaded me on to a stretcher and ran with me into the hospital's malaria treatment room. This was effectively a bare room with about twenty-five high-powered fans surrounding a bed. I was laid on the bed and the fans were turned on at full power while I was also hooked up to an IV drip laced with a cocktail of antibiotics. Later, I was told that if they hadn't been able to control and then reduce my temperature, I'd probably have died within twenty-four hours.

The remarkable thing was that, only a week later I was feeling okay again though the doctors insisted I rest up for another week before rejoining my unit back in the countryside. Luckily, I've never had a recurrence, but, having suffered that serious bout of the infection, I now know why so many Allied soldiers died in the Far East from disease in World War II and why the conflict in Vietnam was so hazardous, even apart from the fighting.

But the challenges and risks were the same for all the other nationalities on UN duty in Cambodia. We served alongside

Royal Navy teams from Britain and Canadian infantry units. Ironically, despite the often painful details of Anglo-French history, there was tremendous professional respect between the Legion and the Royal Navy. But for the Canadians all things French posed certain problems. The biggest off-duty brawls in Cambodia were inevitably between Legionnaires and Canadian troopers. More than a few bars were literally wrecked as Legionnaires and Canadians decided to stage modern re-enactments of the Battle of Quebec.

The six-month tour of duty absolutely flew by. I felt as if I was born to do this. And, when the elections went without any major violence, all the UN forces began to relax and enjoy the wind-down. For me, that meant a one-week period of leave. Initially, I had planned to go with another Legionnaire, an ex-US Marine called Ascot, on a motorcycle tour of Cambodia until a few older Legionnaires, knowing we would go wild and probably get into serious trouble, asked whether we were insane. They strongly advised us to spend the entire time in Thailand at a beach resort like Pattaya or Koh Samui. A few officers, knowing that we were almost broke, even gave us a few dollars to fund our way. I think they were also concerned at the prospect of what we might do running loose in Cambodia. The flights to Thailand were free, the Thai hotels were cheap, the beer was plentiful and the girls, well, they were there too.

Ascot and I ended up in Pattaya, which was ironic given that the resort was initially developed in the 1960s by the Americans for GIs taking leave from the conflict in Vietnam.

Even now, I have only hazy memories of that week. One of the things that I can remember is a ferocious bar brawl between various soldiers, all on leave from Cambodia. Typical of my luck, I ended up scrapping with a US Air Force Corporal who was one of the biggest men I have ever seen. He must have been six foot eight inches in height, but size meant nothing to me because I was a Foreign Legionnaire and he wasn't. It was madness, total insanity – but we were young, we thought we were invulnerable and we were determined to go wild.

Ascot was a great companion and a tough customer. He had joined the Legion after leaving the US Marine Corps. He was another guy who, if you were in a tight corner, you wanted by your side. Typical of how wild our time was in Pattaya was the fact that we met two beautiful Swedish backpackers one night but were so drunk we couldn't remember where we said we would meet them the next night!

Another night, Ascot and I got involved in a street brawl and had to make a run for it when the local Thai police were called. The only way we could escape was by jumping into the surf from a restaurant built on a promenade out over the sea. We then had to swim parallel to the shore before reaching another bar built on stilts out over the water. Climbing the heavy wooden stilts, which were covered with barnacles, to reach the bar veranda left our arms and legs torn and bloody. Even now I cringe to think of the state we must have been in, standing at the bar and ordering drinks while the combination of seawater and blood trickled off us on to the floor.

How Ascot and I ever made it back in one piece from Pattaya I will never know. But I'm sure the Pattaya police, who were tough men in their own way, were all mightily relieved when the Legion's Cambodia tour was over and they didn't have to deal with lunatic Legionnaires on R&R.

Not long after we returned from leave, the REG's tour in Cambodia was over and it was back to France. Typical of the Legion, we were first in and last out on almost every assignment. The 6th REG were charged with airport security and we were the last French force to depart Sihanoukville as the UN mission wound down. But I had completed my first overseas mission with the Legion, and had even earned a commendation for my service while in the boonies. I knew that this would almost certainly speed me on my way to my Caporal's stripes.

Coming back to barracks in Lardoise after leave or overseas duty was always tough. Firstly, you were usually broke for a few days until your back-pay came through, often hung-over and had put on a few extra pounds from eating and sleeping too much. It was enough to make some of the Legion sergeants foam at the mouth with anticipation. For a full week, returning soldiers went through a gruelling fitness regime, then, after six or seven days you were back in shape and the excesses had all been sweated out of your system.

Settling back into Avignon itself wasn't quite so easy. There were always tensions between the Legion and the local Arab population. To be blunt about it, there was mutual loathing on both sides. When we were in town, we had the clubs and bars we frequented – and Legionnaires liked to be left alone. For the most part, Legionnaires and young Arab men stayed away from each other. But add a few drinks to the mix and you got a cocktail that was likely to explode at any time. The natural interest of young men – from both the Legion and Arab populations – in pretty women only served to make things substantially worse.

Usually it was the small things that lit the fuse. When the bulk of the regiment was on assignment away from Avignon, the local Arab lads regarded our bars and nightclubs as fair game. They effectively took over for a few months. And then we arrived back in town from Cambodia, Bosnia or Africa, bristling with energy, cash and hormones. If it was obvious that you were a Legionnaire, you'd probably be refused admission at those clubs now, or, if you were with a woman, there'd be a few smart comments passed about you in front of her.

Not that Legionnaires were subtle about their feelings. After one incident when two Arab doormen on duty at a particular nightclub had allowed two pretty girls into a club and then deliberately refused admission to their two Legionnaire boyfriends, their company colleagues decided to send a rather blunt message – and the next night they threw a smoke grenade into the office used by the security men without, of course, telling them that it was only a smoke grenade and not a fragmentation round.

On another occasion, I was walking down the main street in Avignon with a friend, both of us royally drunk, when we got into a row with two passing Arab guys. Before long, punches were thrown – probably by us – and the four of us squared up to each other. But there were far more Arab men around than Legionnaires and, within minutes, instead of facing two guys we found ourselves battling a mob of about twenty angry Arab men. Eventually, I was knocked to the ground and was on the receiving end of a pretty heavy kicking. All I can remember is my friend grabbing me by the neck and hauling me backwards just before an Arab with a large carving knife drove it down towards the spot where my head had been seconds before. Luckily, the sight of the knife shocked the crowd and people instantly began to disperse when they realised what might have been. After that, no matter how drunk I was, I was pretty careful about where and when I got involved in rows.

For French-born Legionnaires, the tension towards Arabs was on an altogether higher level. Non-French Legionnaires like myself didn't help matters much by offering the opinion that Marseilles was the new capital of Algeria!

Some years later, just weeks before my final tour of duty in Bosnia, the most serious clash between the Legion and the local Arab population occurred. Two Legionnaires got into a dispute with some Arab guys over a woman and it turned nasty. The two soldiers were jumped by an Arab gang and one of the Legionnaires was beaten senseless. He was in hospital for months afterwards. The Legion heard what had happened and decided a message had to be sent out – we protect our own. The area where the gang lived was identified and, that night, Legionnaires moved *en masse* into the district. The flats complex where several of the gang members lived was located and every single flat and bedside was visited by angry soldiers. It was the last time Legionnaires were so blatantly targeted in Avignon during my service.

Otherwise, life back in Avignon was training, manoeuvres and praying for an overseas assignment. The Legion was not a barracks outfit, for two main reasons: the Legion was entirely focused on operational deployments and, to maintain that state of constant readiness, ferocious discipline was maintained while in barracks. In fact, being sent to Cambodia or Bosnia or Djibouti was actually a relief from barracks life. The second reason was that an overseas deployment also meant more money in your wage packet.

Not long after returning from Cambodia, I got my first home leave with the Legion. I'd been a Legionnaire now for a year and I knew that I'd been earmarked for promotion. I'd been told that, within weeks, I would start preparing for my Caporal's training. I also got a lot of my back pay and for the first time in years, I felt I was a wealthy man. I felt so contented with life that I reckoned I could have floated back to Cobh instead of catching the Aer Lingus flight from Charles de Gaulle airport outside Paris.

After confirming my leave with a lot of other Legionnaires from the REG and other units of the Legion Rapid Reaction Force such as the Cavalry and Infantry, I booked a train ticket to Paris from Marseilles on the TGV. There must have been forty or fifty Legionnaires on the train, all heading back to their various homes throughout Europe – and all determined to get ferociously drunk *en route*. The only problem was that the rush-hour TGV was packed and the bar carriage was full to capacity. But one Parisian Legionnaire had smuggled a harmless black grass snake back to base from Cambodia, and, with admirable foresight, had decided to pack the reptile in his kit bag. With a knowing smile, he told us to wait outside the bar carriage while he emptied it for us. Less than thirty seconds after he discreetly dropped the large snake on the floor, the carriage erupted in a wail of screams and howls. There was a mass stampede for the door and, less than two minutes later, we had the entire carriage to ourselves.

After reassuring the terrified bar workers that the snake was in fact harmless, we proceeded to drink our way all the way to Paris. By the time the train pulled into the station the bar had been emptied – again! I literally fell off the TGV in Paris and then hit town for another twenty-four hours before catching my flight home to Cork. It was the same once I got home. I met my parents for the first time in a year and, after assuring them that I was fine, I hit the town with mates I've had all my life. The three weeks passed in a blur of parties, discos and pub sessions. I cut loose in Cobh like a twenty-one-year-old whirlwind. I was immensely proud of myself. I'd not only proven that I'd got what it takes to make a soldier, I was a Legionnaire, the toughest of them all.

The thought of not returning to the REG in Avignon never entered my head. In fact, it was almost a relief to get back and escape the party lifestyle that had made my Cobh stay such a blur. The pain of training and barracks life was eased by the fact that I was earmarked for promotion to Caporal. I was thrilled, even though it meant I had to return to Castel to complete my Caporal's course.

That's what I focused on throughout 1994 – getting my stripes and the additional qualifications I needed in mines and explosives. Unfortunately, it was during this period that I met an officer who would have a huge impact on my military career and help turn my dream into a nightmare. During that summer we were on manoeuvres in the Larde-che area – it was part of our amphibious training and we were doing white-water kayaking on descents of a ravine. Of course, we were carrying almost our full kit, including our FAMAS assault rifles. The officer in charge, Lieutenant Guyot, presumed that we had been shown the correct proce-dure for tying the rifle to your leg so that, even if the boat overturned, the weapon didn't get lost. Unfortunately, no-one had shown us the drill and, when the boat directly behind me overturned after it hit a rock, two Legionnaires were thrown overboard and their rifles were lost in the deep swirling waters of the ravine.

The ravine is famed amongst water sports enthusiasts for its white-water rapids, and, in the circumstances, we should have been grateful that the two Legionnaires got out of the river alive and uninjured. But in the Legion for every action there is consequence: the rifles were lost not through an acci-dent but because precise procedures were not followed, and that had to be somebody's fault. And one of the things I've learned in life is that shit usually flows downhill.

The entire unit was immediately ordered to commence a search and, after several dives, we managed to locate one of the missing rifles at the bottom of a deep rock pool. But, try as we might, we just couldn't find the other FAMAS. At least one mile of the entire river was searched – I think we got to know almost every individual rock in the whole ravine. But there was still no sign of the FAMAS. Eventually it had to be reported missing to the base armourer in Avignon and, in turn, the word went to military headquarters in Aubagne and Paris. There was hell to pay. Despite the search we had already conducted, for almost a month other Legion units were sent to the ravine to search every nook and cranny for the rifle. To make matters worse, word leaked out that a Legion rifle had been lost and a *gendarme* had to be placed on duty in the area to prevent weapon hunters from moving in.

The two Legionnaires involved were immediately transferred out of the section, and, a short time later, Lt Guyot also left. He was French and had apparently wanted to join the Legion in the ranks. However, his family proposed a wiser course and he attended the officer school at St Cyr. He did exceptionally well and was one of the upper cadre selected for the Legion. Initially I found him a great guy, the kind of officer you pray you get to serve with. During his first few months with us at Avignon he would even join the NCOs and ordinary Legionnaires for a few beers. Sadly, after the incident in Lardeche things would change and, tragically for me, our paths were destined to cross again and I somehow became associated in his mind with the incident.

I was one of only five guys still left in the section after the fall-out from the lost rifle incident. There are few things in the Legion as disgraceful as losing your weapon, particularly when it is peacetime and during routine manoeuvres. But it also meant that I was now one of the more experienced

Legionnaires in the section and, with my Caporal and hope-
fully sergeant stripes on the horizon, all was looking very
rosy.

I also knew that another overseas assignment was likely
quite soon because the Legion had never been this stretched
for overseas missions since the old days of the French colonial
empire. There were permanent Legion detachments in Djibouti in Africa, Guyana in South America, Mayotte in Mada-
gascar and the French Pacific islands where the nuclear test
sites were located. Occasionally the Legion was also assigned
to Chad and the Congo in Central Africa. Given the tours of
duty and rotation requirements, it meant that Legionnaires
were spending less and less time in barracks and more time
overseas. That was a fact that suited me perfectly.

Another fact that wasn't lost on anyone in the Legion was
that the crisis in the Balkans was getting worse daily. The TV
headlines almost every night in France were dominated by
the bitter clashes between Croats, Serbs and Muslims. The
siege of Sarajevo was also dominating world headlines. Of
all the assignments open to the REG, I reckoned that the Balkans would undoubtedly be the most demanding. Not for
the first time I was proved correct.

5

HELL ON EARTH – SARAJEVO

There are some places on earth and you just have to wonder whether they've been cursed. I'm not a great believer in omens of bad luck, but Sarajevo has had enough trouble over the years to make you think twice. World War I started here with two gunshots, and now, eighty years later, the locals were still intent on butchering each other.

The word had gone out in Avignon over Christmas 1994 that we were up for an overseas assignment and, within days, every Legionnaire knew that it was going to be the Balkans. The Paris authorities had been one of the first to provide troops under a United Nations mandate in 1995 which was aimed at trying to defuse tensions and end the conflict which had made household names of Radovan Karadic and Ratko Mladic. Little did I know as I prepared my kit just how personal my contact with Mr Karadic would be.

The 1st Company of the 6th REG flew out in early January and my unit was assigned to Sarajevo which was still under siege from Serb forces. The tensions were running so high that our aircraft landed at the airport under full combat conditions. It raced into the terminal under full power from the runway and we had to run from the aircraft loading ramp into the safety of the terminal building. As I would quickly learn, Sarajevo was being turned into hell on earth by snipers – no-one was safe, from heavily armed Foreign Legionnaires right down to innocent mothers and children.

After being processed at the airport, the Legion units were sent to the assignment areas. My unit of the REG was to be attached to a large UN base at Scandarija, which was located within the city. For the next five months this old complex would be my home. It was a huge sports complex in the city suburbs that had been built when Sarajevo hosted the Winter Olympics in 1984. But there had been some change: now, Scandarija was surrounded by security perimeters, and sandbag revetments marked out guard posts. There was barbed wire everywhere.

The trip from the airport to Scandarija left us in no doubt as to what we faced. The landscape reminded me of pictures of Stalingrad, Berlin or Warsaw after World War II. What had obviously once been a very beautiful city was now little more than rubble and the shells of scarred buildings. There was evidence of heavy artillery and mortar fire everywhere. Some buildings were so heavily hit by bullets that they looked like concrete sieves. Every person we saw was hunched over and hurrying to get to safety. No-one seemed happy about being out in the open. And on every major intersection there was the tell-tale sign of heavily armed militia groups. You could have been forgiven for thinking there was a smell of absolute evil on the wind. Sarajevo looked like a city that had been abandoned by any feelings of decency, honour and nobility.

Luckily, we were told that we would mostly be working within the French battalion's area, which meant working alongside units from the French Marines. During manoeuvres I had learned that these were great guys and ranked right alongside the crack units of the world like the Legion, the Royal Marines and the US Marines. Given our specialised training we would be available to other UN forces if required, though, as it transpired, we only undertook one such mission for a British Army detachment.

Because of my explosives training, I would be responsible for mine clearance duties. This meant clearing access routes of anti-personnel and anti-tank mines planted by the various militia groups. But we would have to secure our own perimeters through planting swathes of anti-personnel mines. Our Legion units would also have to support anti-sniping patrols, and this was a thankless, heartbreaking task.

Just two kilometres from our base at Scandarija was the so-called 'Sniper Alley'. This was the main road out of Sarajevo and it was plagued by snipers who fired from multistorey buildings on each side. The butchery along this road had dominated headlines all over the world for more than twelve months. Not far from this road was the open-air market where a surprise Serb mortar attack had massacred civilians desperate to buy fresh meat and vegetables. The grim reality here in Sarajevo was that no-one was safe –

man, woman or child. Anything that moved was deemed a fair target, even if you were wearing the blue beret of a UN peacekeeper.

The situation on the ground was complicated to the point of being mind-boggling. Specific parts of the city and its suburbs were controlled by different ethnic factions but there could be tensions even within those factions. The length and savagery of the siege had taken its toll and, throughout Sarajevo, there was a mood of gloom and despair. Everyone feared the snipers and everyone was terrified of the militia gangs, several of whom were little more than death squads. The black market was rife and almost everything and everyone had a price. As usual, it was the ordinary working people who bore the brunt of the violence. When we arrived in Scandarija I was astounded to see that the huge underground carparks were full of Mercedes cars, BMWs and Volvos belonging to the Sarajevo elite – these luxury cars were being carefully stored until better days returned. The location of a UN base around the storage area was an added bonus.

Luckily for me, I had been assigned to a command unit operated by the 6th REG. This meant I was working alongside a lieutenant, a master-sergeant, two Caporals and two Legionnaires, usually with two VAB armoured personnel carriers (APCs) at our disposal. The VAB was a Renault-built APC which dated back to the early 1970s and was renowned for its ruggedness. It were powered by a MAN diesel engine which often made conversation inside the VAB very difficult. But it were well armoured, could go almost anywhere and was amphibious. The VAB was even equipped with a Nuclear Biological and Chemical (NBC) defence system. But it was lightly armed, usually with only a squad-heavy machine-gun. We didn't have any specific duties but were to float between whatever urgent missions were required. Other units weren't so lucky and spent almost their entire time in Sarajevo on Sniper Alley – I only did occasional duty there. In the bars and cafés these duties were grimly referred to as 'sniper bait'.

Such anti-sniper missions involved small UN teams being assigned to more than twenty observation posts along the entire length of Sniper Alley. Troops were told to scan buildings constantly with high-powered binoculars for any sign of sniper activity and, if engaged, report immediately back to their battalion HQ. The theory was that the battalion would then contact UN headquarters and, after a quick consultation, the post would be advised about what action to take – which worked fine if the Legionnaire or trooper was still alive and hadn't already had his head, complete with his UN blue beret, blown off by the sniper.

Sniper Alley taught us that the UN motto was: 'Do, but don't.' It was like trying to fight a war with both hands tied behind your back. We knew it – but, tragically, so did the Serbs, the Bosnian Muslims and the Croats. In Sarajevo our every move was scrutinised and judged; at least in the countryside you would have a small amount of unit discretion, but in Sarajevo we often felt that we were alive only because the militia groups didn't want the hassle of killing us.

But there were exceptions. Once, I was assigned to a mine-laying detail on a roadway parallel to Sniper Alley where we had an observation outpost called 'Sierra November'. The UN had strict rules about such missions. The location of all our mines had to be recorded and the direction of the trip-wires carefully noted. The Serbs, for their part, warned that any offensive action by the UN would be met with force. We regarded the mines as defensive, but no-one knew what the Serbs would think about that. I was working on my hands and knees in an open area with Master-Sergeant Leurs beside me when I heard the tell-tale whizz of two 7.62mm Kalashnikov rounds passing within inches of us. The dust from one bullet as it impacted on the ground actually hit my face. Instantly, we both broke for cover and tried to shield ourselves from the wrecked multi-storey building where we judged the fire had come from. This building was between us and Sniper Alley. Less than two minutes later, we heard two fresh rounds being fired, but they seemed to have been fired in a different direction. Slowly and carefully we made our way back to the safety of our vehicle and reported back to base.

But what we hadn't realised was that the sniper, having missed in his overly ambitious attempt to kill both myself and Master-Sergeant Leurs with a single salvo, had run into another apartment facing on to Sniper Alley and opened up again. It so happened that a regular French Army colonel was giving an interview live on French TV when the sniper fired. An unfortunate army private was working on a heavy-loader in the background, lifting concrete panels on to mountings in an attempt to offer protection from snipers during the interview. The first 7.62mm round hit the windscreen of his tractor and, instead of getting down on the floor of the cab, the driver opened the door and ran for cover. The second 7.62mm round caught him in the throat – all live on French TV.

One week later I was driving a VAB armoured personnel carrier at the same spot, dropping my team at Sierra November. As I turned the VAB in the roadway, I spotted some militia personnel on the road beginning to engage targets in a high-rise building directly behind me. Their rounds were screaming over the top of the VAB – and I realised they were using my VAB as cover. Then I watched as one militia man raised a rocket-propelled grenade (RPG) rifle and fired. The explosive round screamed over my VAB, missing the windscreen literally by inches. I could actually see the flames from the rocket as it passed; I followed it and saw it explode in a building behind me. I then realised that my VAB was being used as a shield by the Muslim militia to fire at their Serb rivals. I knew I was damned if I stayed where I was so I put the VAB into gear and roared away. This was a normal day at the office in Sarajevo.

It's hard to believe that, in the midst of all this carnage, you could still manage to find something to smile about. But one day I was assigned to Spot 18, Sniper Alley, on anti-sniper duty and, as I got out of the VAB armoured personnel carrier, I spotted a sign over a small local bar which said '*Céad Míle Fáilte*'! I couldn't believe my eyes and went over to investigate. Sure enough, it was an Irish bar being run by two young Sarajevan ladies. To be honest, it was a typical Balkan-style café which had been turned into an Irish bar by sticking a shamrock on the window and putting up a few green ribbons. Outside the front door, the *Céad Míle Fáilte* sign even had bullet holes in it. But we had found on unofficial Legion canteen – and, for the remainder of our tour this was where we took a break from the horrors of Sniper Alley. In future, if duties on Sniper Alley were being assigned, I generally took Spot 18 – not that you could get Guinness or Beamish or anything remotely Irish to drink – it was Sligovitsa or other such local fruit brandies, but it was better than the syrupy coffee available in the UN mess.

Other Legionnaires opted for a different kind of activity. One evening I was on guard duty with another Legionnaire at Scandarija when a small militia unit, either through boredom or some local feud, decided to open fire on us. I was in the perimeter command post and could hear the 7.62mm shells whacking into the sandbags around us. As my Legionnaire companion was not to be seen anywhere, an overly enthusiastic French Marine Sergeant decided that someone had to check all our perimeter outposts. Needless to say, I thought the idea was sheer lunacy. I looked down the tunnel leading from our command post and could see the bullets whizzing by. However, I was ordered to do a reconnaissance, and, despite my misgivings, realised I had to fulfil my orders. I sprinted from the command-post tunnel to the open-air parking space where our heavy trucks were stored. I stood in front of a truck, taking cover, and in the darkness I could hear the rounds crashing behind me and ripping into the canvas cover of the truck.

I checked all the outposts and bunkers but couldn't find

the Legionnaire. I hissed his name into the darkness several times and got no answer. I checked the post and he wasn't there. Just as I was about to raise the alarm, I heard a grunting sound coming from the outer perimeter near the barbed wire. As I crept over to investigate, one of our trip-wire flares was triggered – and I still can't believe what I saw. Beyond the barbed wire, a militia figure was standing and carrying a rocket-propelled grenade (RPG) which was raised to his shoulder. I instantly scrambled clear of the bunker expecting at any second to hear the 'whizz' of the incoming RPG round. At the same time, I was shouting the missing Legionnaire's name.

But no impact came – the militia man had obviously had second thoughts and fled. Then, as I shouted the Legionnaire's name again, I finally got a reply. I ran to the other side of the bunker and there he was, desperately trying to pull up his trousers while a local prostitute, on the other side of the barbed wire, was scrambling to get away.

The late-night liaison came at a price for that Legionnaire. As Caporal, I had to choose a Legionnaire to raise the Tricolour the next morning and, because the flagpole was in an elevated position by the boundary wall and open to sniper fire, it was a duty everyone hated. I figured if the love-sick Legionnaire could drop his trousers so quickly, raising a flag fast shouldn't be too much trouble for him!

There were unexpected consequences too. The next day, I bluntly warned a contact in the local militia that I didn't want to see any hookers touting for business around our compound. I meant it as safety advice, both for the ladies and our troopers. It was far too dangerous for everyone involved, given the regular militia attacks. But I later saw the terrible result of my warning when the prostitutes turned up anyway, but battered and bruised after a savage beating by the militia. Such was life in Sarajevo.

About two months into our tour, I was on command team duties one day when word came into base that one of our heavy trucks had broken down in the mountains some 15km from Sarajevo. It had been on a humanitarian mission assisting with a clean-up and the restoration of utilities in a village behind the Serb lines. A detail of one Sergeant and two mechanics was assigned to repair the heavy-duty BENNE truck and get it back to base. As the Lieutenant I was assigned to was off on other duties, I offered to accompany the rescue team and man the .50 calibre heavy machine-gun mounted in the rear of the VAB jeep. We had to move fast because at 8.00pm curfew began and all road traffic stopped. Anything that moved after that was likely to get fired at.

We passed out of the city and went through the Serb checkpoints with little difficulty. We reached the truck and, within an hour, the mechanics had the BENNE tipper truck up and running. Unfortunately, we were still in the middle of the severe Balkan winter and the roads were sheeted in snow and ice. We had only travelled about four kilometres back towards Sarajevo when the driver lost control of the truck and it ploughed off the road. A quick inspection confirmed that there was no way we could get it free in time to make it back to the city before curfew. Now we had to figure out how to overnight behind Serb lines.

On parade: no matter what the rigours of training are, the Legion is obsessed with parade-ground reviews and immaculate dress. A Legion joke goes that the two keys to being a good Legionnaire are being able to shoot straight and iron your dress uniforms immaculately.

It's a tough, painful climb over the Mont Louis scaling wall – particularly for those Legionnaires still at the bottom. I'm on the right here, about to clamber to the top on a comrade's shoulders. It's all about team-work, and fighting through the pain barrier together.

The Legion took a malevolent pleasure in its tough training regime. Here a group from our regiment tackle an obstacle course at Mont Louis, the infamous commando training centre in France that is reckoned to be one of the toughest in the world. I'm at the bottom left, determination and pain etched all over my face.

Mont Louis takes no prisoners. Here I tackle the feared 'death slide'. The two rules at Mont Louis are that you never quit – and you never look down!

The Legion's love of formality and ritual is evident at this graduation parade.

They say that you're not a real Legionnaire unles you've done time in a Legion prison! Here I share jokes with fellow cellmates.

CLOCKWISE:

The standard Legion beret. The parachute regiments liked to adorn theirs with the para wings.

A collection of the medals and service awards I earned in Cambodia and Bosnia.

A standard Legion shoulder 'Flash' designed to make our nationality and unit more identifiable.

My graduation year; I am second from right, standing.

My Legion ID.

A backbreaking introduction to UN duties in Cambodia. We were assigned to reconstruction duties, and that meant repairing bridges, culverts and roads in the baking heat and searing humidity of the South East Asian summer. Here my Legion crew works on repairing a road bridge. Note the lack of any heavy machinery.

Our unit enjoys a well-earned rest complete with a few beers after a gruelling Cambodian patrol. I'm at the rear, looking for an extra beer. The NCO with the epaulettes and moustache in the foreground is Master-Sergeant Gercer, one of the toughest men I've ever served with and a living legend within the Legion.

I pose beside a UN-marked VAV that bears all the hallmarks of conflict: the paint chips missing from the armoured hull where sniper rounds had hit. The vehicle on the right is an engineer-support vehicle capable of removing debris, tackling roadblocks and even recovering stranded trucks and jeeps. Note the damaged condition of the buildings overlooking our compound.

A close-up view of death and injury. These anti-personnel mines took a terrible toll on ordinary Bosnian civilians. Note how rusted the mines are due to the severity of the Balkan weather. This made the mines even more dangerous and unpredictable.

The first thing we had to do was take a holding position with the convoy and inform the Serb authorities of our plight. There was a police station down the road and the Sergeant drove me down to explain to the police what had happened. He then returned to the truck and mechanics, and I set about trying to arrange for a Serb policeman to travel back to our convoy and stay with us until we were ready to move out the following morning. I knew the Serbs would be suspicious of us and I would have to try and defuse the situation in the police station as quickly as I could. So I walked into the station, said a friendly 'Hello' to all the police and militia personnel and walked right across the room before placing my FAMAS rifle on its tripod against the wall. I didn't want them to feel threatened by the rifle, but they weren't going to take the weapon off me either.

As I turned around, one of the Serb policemen spoke to me in reasonable English. I explained about our accident and the fact that we would have to stay put overnight. He knew I was Foreign Legion and I then explained that I was Irish. After chatting for a few more minutes he introduced himself as Lillian and said that he had once been on the Yugoslav Army kung fu and karate teams. I smiled back, trying to be polite. He then turned to the other policemen and said something in Serbo-Croat. The men laughed and I thought: Fuck this, they're not going to laugh at me or the Legion. So I turned to the Serb, eye-balled him and asked him what was so funny. My voice left no-one in any doubt that I didn't give a shit who they were. I was Legion – and they weren't going to intimidate me. The tone of my voice instantly changed the atmosphere in the room. I knew I wasn't going to back down. The guy said that his kung fu training was so good he reckoned he could split one of my body-armour plates with a single kick.

The body armour was made of Kevlar and was designed to offer protection from rifle calibres as high as 7.62mm. I thought to myself, if a Kalashnikov round can't get through this Kevlar, then no Serb is going to kung fu it in half either. Not wanting to be intimidated, I grinned and told him to give it a try. Seconds later, I had been slammed back against the wall with the force of a drop-kick to my chest plate. The other guys thought it was hilarious. Despite the weight of my body armour, I had the wind knocked out of me – but I still wasn't going to back down, so I invited Lillian to try again. For the next ten minutes, the Serb policemen were rolling around the ground in hysterics while Lillian treated me like a kung fu target. In fairness to him, every kick was aimed at my chest armour.

Finally, they lost interest in the game, to my relief. I was never so grateful for body armour. But the men had clearly decided that I was okay. One got a chair and indicated that I should sit down. No-one mentioned my rifle which remained untouched by the rear wall. And then Lillian produced a bottle of local brandy and poured two glasses. For the next two hours, we drank and toasted each other. They all referred to me as 'Irish' and, as the brandy flowed, Lillian proposed that he should personally take me hostage. This would allow him to take me back to his home where, he promised, his wife and family would fête me like visiting royalty. The ultimate Serb tribute was that he offered to slaughter and roast a pig in my honour, as much to show how welcome I was as to offend the Bosnian Muslims who were his neighbours. Ironically, he also explained that he got a lot of slagging from his fellow Serbs over his name, with 'Lillian', apparently, having been derived from the *fleurs-de-lis* that were on the Bosnian flag. He also said one of his relatives had a fine cellar of local wines and brandies and proposed a drinking marathon after the barbecue.

I politely declined though the prospect of a few drinks was certainly enticing. No, I knew I had to try and get back to the Legion trucks as soon as I could. And then my world was turned completely upside down. I heard the sound of vehicles approaching outside and one of the policemen stuck his head inside the office and shouted something. Instantly, the atmosphere changed and the policemen who had been relaxed and amiable suddenly were businesslike and brisk.

A minute later, the door opened again and a large group of men walked in. It must have been close to midnight. Several were in uniform and were evidently senior Bosnian Serb commanders. But they all seemed to defer to a well-dressed man with grey hair in the centre. I looked in astonishment as I realised I was staring at Radovan Karadic himself. On either side of Karadic stood two heavily-built bodyguards and a middle-aged woman who acted as his interpreter and personal assistant. Instantly, the bodyguards' eyes fixed on me. I glanced to my rear at my FAMAS rifle. The weapon was at least five feet away, and both bodyguards recognised the look for what it was. The two men opened their overcoats and rested their hands quite deliberately on their side-arms. I saw that they carried Yugoslav-made Zastava 9mm automatics, a local version of the Red Army's Tokarev.

Both these guys were pros and I instantly knew that I wouldn't make it to my rifle. So I just sank back into my seat. They stared at me as a weasel eyes a rabbit. Karadic, oblivious to this silent exchange, had been interrogating the local police about me. He then turned and walked over to me, extending his hand. I shook it without thought and then watched in growing alarm as he turned and walked to the door, gesturing at me to follow. I picked up my rifle and a bodyguard instantly positioned himself directly behind me. I had no choice but to follow.

Outside the police station was a large S class Mercedes with two other escort vehicles, one of which was a paramilitary jeep. The Mercedes looked like it had been armoured. I was guided into the rear of the black saloon and Karadic sat in front of me as the car swept away. As we turned off the main road and made up a mountain track, two words formed in my mind: shallow grave. No-one would ever find my body, let alone discover what had happened to me. I wouldn't be the first to disappear without trace in Bosnia.

We travelled for about thirty minutes before pulling into an isolated farmhouse. I was motioned to get out of the car and was tensing for what was about to come. But then I saw a woman emerging from the house and coming over to embrace Karadic. He gestured to me and I smiled back at the woman. She disappeared back into the house only to re-emerge and hand me a home-made fruit cake. It was another twist on the emotional roller-coaster I'd been on since mid-afternoon. I waited outside while Karadic spent some time at the house – clearly they were either relatives or friends. When he emerged, he smiled at me again and our little convoy disappeared back down the road. Less than thirty minutes later, I was dropped near our stranded convoy, and felt myself the luckiest man on the planet to be still alive.

Incredibly, the lads were asleep in the rear of the VAB because a regular French Army security unit had been despatched to the site. It was now well after 3.00am and it was absolutely freezing out in the open. I got into the cab of the VAB and held on to my rifle. I was too pumped up with adrenalin to sleep. So I thought I'd keep guard until the morning. I stayed like that for a couple of hours until I thought I heard a noise. Earlier I had spotted what I thought were the lights of a vehicle coming up the mountainside but they had disappeared. Instantly, I got out of the VAB and listened carefully. I heard the noise again – it sounded ominously like the click of a Kalashnikov rifle swinging on its harness against spare magazine clips. I'd got to know the sound from watching the militia groups march and patrol around our camp. I slowly lowered my FAMAS and stared at the winter gloom in front of the jeep from where the noise had come.

The sound of crunching on mud, ice and snow confirmed my worst fears. Armed men were approaching our convoy – for what other reason than to take us out? Then, just as I was about to get into a firing position, I was startled by a shout from the darkness. I couldn't make out what was being shouted at first but then, despite the Serb accent, I realised that someone was calling: 'Irish, Irish.' I called back that it was okay to come forward and was stunned to see four Serb policemen walk up to the VAB and hand me a parcel containing water, bread and some tins of sardines. They explained that Lillian had sent his regards. I thanked them and they immediately walked off into the darkness. I got back into the jeep not knowing whether to feel relieved, shocked or confused. I was on a roller-coaster of emotion and adrenalin. The next morning we drove back to Sarajevo as if we were returning from a mountain picnic.

What struck me most about Bosnia was the emotional pendulum that we constantly operated under. You could go from fearing for your life one minute to having a laugh the next minute with the guy you thought was going to kill you. But it slowly took its toll on our nerves, and the mounting levels of bureaucratic bullshit with the UN – and the Legion – only made things worse.

In the middle of our anti-sniper campaign and not long after my Serb experience in the mountains, a regular French Army Colonel was on an inspection tour of our Scandarija base. As is the norm, all the French units at the base turned out on the parade ground for a formal review. But the Colonel spotted that while the regular army units were singing 'La Marseillaise' with gusto, the Legion members were stiff at attention but silent, steadfastly failing to sing the legendary national anthem penned in 1792 by an army engineer, Claude Rouget de Lisle.

The Colonel went ballistic and, within minutes, had demanded an explanation of the Legion commanders. By tradition, the Legion will honour 'La Marseillaise' by standing to attention but, because it is a foreign force, it does not normally sing the song on review. It was a tradition that had largely been respected for decades. Of course the Colonel chose to ignore this and the fact that, for almost 170 years, the Legion had been dying for 'La Marseillaise' everywhere from Mexico to Algeria and from Africa to South-East Asia. No, he wanted the song sung by us Legionnaires and that suddenly assumed greater importance than anything else we were doing in God-forsaken Sarajevo.

Our Legion captain told us we had no choice but to sing the anthem, our lieutenants warned that we had to do it and the Sergeant-Majors were almost foaming at the mouth trying to get the ordinary Legionnaires to learn the song in order to fulfil their orders. The failure to do so would result in disciplinary action and, privately, we were assured that our lives would be made hell if we didn't comply. But the Sergeants, Corporals and ordinary Legionnaires felt that a Legion tradition was being trampled upon – it went to ground level to stand up to this bullshit. We were *not* the regular French Army but the senior officers were insisting that we be treated like them despite the fact that we got the most thankless jobs and often the worst equipment. It left many NCOs deeply disillusioned. While I didn't know it at the time, this type of increasing bureaucracy was one of the many small things that sparked the exodus of some of the Legion's most experienced personnel. When the next review came along, I steadfastly stayed silent in the rear ranks throughout 'La Marseillaise'. I wasn't the only one.

Given all this bullshit, it's not hard to see how Legion-naires needed to let off some steam. One night, four of us decided to break barracks and head into town. This was strictly against orders, but, once you weren't caught and didn't cause trouble, people were prepared to turn a blind eye. Myself and three Legionnaires from my company – Tabone, Barry and Sikorski – set off looking for any sign of nightlife in the city.

Along a semi-wrecked street we spotted a club in the base-ment of a multi-storey building. As a precaution we had taken off all our Legion badges and UN insignia. We wore plain combats and, from a distance, we could have been from any army or militia group. The club, incredibly, was being run by this young lad who was about sixteen or seventeen years old. As we ordered a few drinks, he instantly realised we were with the UN and began to talk to us about the black market. He was able to offer us anything we wanted from booze to women – all he wanted were the consumer items that every household depends on: corn-flakes, branded cigarettes and chocolate.

Sikorski, who had already earned a reputation as a bit of a lunatic, opted out of the conversation and went on to the dance floor. As the music blared loudly he began to dance and shout in Polish, and, in less than twenty minutes, I noticed that the young nightclub owner was losing interest in our conversation and casting nervous looks over at Sikor-ski. His antics on the dance floor were more like a tribal war ritual than disco dancing. I also saw, out of the corner of my eye, two young men, both in fatigues, leave the club.

I thought no more about it until suddenly a large number of uniformed and armed militia men started to arrive. I'd had a few drinks but I was still sober enough to recognise that several of these guys weren't just ordinary militia – they were part of the so-called paramilitary brigades that were associated with some of the worst atrocities in the city. They were not the kind of guys to piss off unless you had their face in the sights of your FAMAS. Sikorski was still dancing like a wild animal and he was now the only person on the dance floor. That wasn't surprising because he was screaming and shouting in Polish in unison with the music and, far from being impressed, any woman would have run a mile from him.

I noted that the mood in the club had become very tense and, with Tabone and Barry, I decided it was time to get the hell out of here. The young club owner came over to us and said that Sikorski had been claiming on the dance-floor, in Polish, that the Serbs were going to slaughter everyone. He was roaring out his hatred of Muslims. In no uncertain terms, the club owner told us to get out – and get out fast. We walked over and grabbed Sikorski and dragged him towards the exit. Incredibly, the Pole insisted on stirring it up. He almost seemed to be looking forward to trouble. I told him to 'Shut the fuck up; we're out the gap here.' The young manager of the club, who had obviously given up any prospect of doing black-market business with us, came over and hissed at me: 'You're all going to die if you stay here. Get out now.'

We finally got to the exit door and I noticed that every

militia man in the club was watching us. We made it through the door and raced up the stairs. The instant we cleared the top step a volley of shots followed us from below. The four of us set an Olympic pace as we sprinted down the street – and then had to dodge between buildings on our way back to base as we tried to avoid militia patrols and our own quick-response team which had been activated after the gunfire in the club. It was our one and only night on the town in Sarajevo.

But it wasn't the last time the mad Pole caused us problems. Shortly after our narrow escape in the club, Sikorski went AWOL in Sarajevo. The word came back to the Legion base that he had linked up with a particularly notorious Serb militia group. This was a major embarrassment to the Legion. Legionnaires had been going AWOL for well over a hundred years, but they didn't usually go and join the forces the Legion was effectively working against. Discreetly, contacts were made with the various militia groups and a huge ransom in fuel was offered for the safe return of the missing Legionnaire.

Sikorski's new friends obviously valued petrol more highly than they did either his military skills or lunatic company. Shortly after the ransom was paid, the Pole was found by a patrol, lying by the roadside not far from our base. He had been hog-tied after being beaten almost senseless. And I reckon that the beating was the easy part of what now faced him. He was taken into military custody, flown back to France and we never heard of him again. The rumours abounded within the Legion but we never definitively found out what happened. There was little doubt but that he faced a military court. And given all that had occurred in Bosnia I don't think they were likely to have been too lenient or understanding of his emotional urges.

But there was also another side to the Legion apart from the combat-orientated, 'march or die', brawling soldiers. Within days of arriving at Scandarija we realised that, for the poor, elderly and sick, Sarajevo was a living nightmare. They had little food, they were in fear for their lives almost twenty-four hours a day and faced into the brutal Balkan winter with little or no heating and fuel. Without hesitation, the entire Legion detachment decided to do something. We instantly worked out a very generous allocation of food-stuffs for our canteen which would be supplied via the French command and the UN. We then pared our actual requirements down to the bare minimum and the balance we secretly transferred to the largest orphanage in our area of operations.

We certainly did our best for those kids. Legionnaires raided military gyms and sports grounds for footballs, shorts and jerseys. Unwanted civilian equipment like torches, batteries and even watches were put aside for the orphanage's use. And when we rotated back to France after our tour of duty, our replacements were handed the respon-sibility of looking after the kids.

When people talk about the casualties of war I think about those orphans. They were only kids and yet they were paying the full price for a war between Bosnians, Serbs and Croats that had been raging since the Middle Ages. For me, the madness of mankind is perfectly evident in the Balkans. Yugoslavia had been a very successful state, people had a reasonable standard of living and it was even developed enough to host the Winter Olympics. Yet because of ancient national and religious hatreds, the whole thing was thrown away. And the price was being paid by orphans and widows.

The hatred had burned so deep that there was little that was sacred left in the Balkans. Our Legion detachment had been assigned to try and clear a minefield outside Sarajevo which was wedged right between the Bosnian and Serb front lines. The minefield had been laid in an ancient cemetery which was regularly visited by families to mourn relatives. As one of the Legion specialists, I was checking for mines within the cemetery when I looked up and, in horror, realised that claymore anti-personnel mines were dangling from the branches of the trees. If the mines had been triggered during a large funeral, the mourners would have been cut to pieces as surely as if they had been put through a mincing machine. When we reported what we had seen, a unit attended the cemetery and noted that every movement was watched through sniperscopes by Serb militia very close by – less than twenty metres. In Bosnia, everyone was a target.

On another convoy escort mission, I was determined to do something to stop the looting and to show that we were *La Legion,* not boy scouts. As we approaching the militia checkpoint, there was a mad scramble of uniformed personnel on to the back of our trucks to unload some cargo. I ordered the driver to reverse our VAB armoured personnel carrier into the lorry and the force of the impact threw most of the would-be-looters from the truck. They angrily screamed at us in the local dialect, at which point I grabbed a .50 calibre heavy machine-gun and levelled it at the militia. There are few things in life to concentrate your mind like the sound of a .50 calibre being cocked, and, as the bullet entered the chamber, the mob instantly went silent.

My Sergeant was hissing at me to calm down but, to be honest, I just didn't care. I was sick of standing by and watching ordinary people suffer and was rapidly losing all respect for what the UN stood for. Our mission in Bosnia was a joke – as far as I was concerned we were helping no-one. We knew it, and so did the Serbs, the Bosnians and the Croats. We were glorified aid workers. We had been trained to take military action, we knew the action that was required and yet we were told to sit on our hands. And while we stood by, innocent people died. I was sick of all the bullshit – the constant walking-on-eggshells when it came to dealing with the militias, many of whom were little more than street thugs and common criminals. I released the .50 calibre and jumped down to the ground. I had my FAMAS held across my chest and I strode into the militia hut, determined to show them that if they wanted hassle, I was ready for them to bring it on. I eyeballed the militia men in the hut while gently fingering my FAMAS and I think they knew I meant business. Not one of them met my gaze or tried to interfere again with our convoy.

We weren't sorry to leave Sarajevo after our tour, though none of us was looking forward to the mind-numbing routine of barrack life back at Avignon. But, after a while, Sarajevo just seemed to suck the life out of you. It seemed to be a magnet for the very worst in human behaviour and there was little nobility left. To be honest, most of us Legionnaires were shocked that what we saw in front of our eyes could be happening in a modern European country. Personally, I lost a huge chunk of my faith in humanity and I couldn't believe that European powers could sit idly by and watch such carnage literally on their doorsteps. What made it worse was that we seemed powerless to help or even to intervene.

Back at Avignon, life quickly settled into the Legion's idea of barrack routine – and that was enough to make us think twice about which was worse, *corvée* and the parade ground or Sarajevo. Mercifully, our barrack time between overseas tours was interrupted by the kind of duty that most Legionnaires can only dream about – an assignment in France that is far from any Legion base.

In the months before we'd returned from Sarajevo, there was mounting concern within France over the potential threat posed by Islamic terrorists and, in particular, whether fundamentalist groups were trying to infiltrate cells in mainland France. It was a fear that was all too real to the Paris authorities. France had fought – and lost – a vicious war to hold on to Algeria. The aftermath of that resulted in tens of thousands of North African landowners returning to France, some of whom had never seen the country of their parents or grandparents. Thousands were Muslim and, like the British immigration of the 1950s and 1960s, the cultural fabric of some parts of France began, irrevocably, to change.

This was particularly true in parts of the south, including

Provence, Languedoc, Roussillon and even parts of the Basque country. The Paris authorities, in the light of mounting fears, became increasingly concerned about security on France's borders, especially the border with Spain by the Pyrenees. It was decided to bolster the regular border police with units drawn from the Legion, and elements of the 6th REG were selected for patrol duties on parts of the border.

I thought this assignment was just brilliant. We could escape the drudgery of barrack life, stay in France and work in an area where it wasn't impossible to get a few beers! Not being shot at was an even greater bonus. We were on manoeuvres just outside Avignon when we got the alert. So, with my unit, I kitted up and headed south-west towards the Pyrenees which was already well known to us from the famed Mont Louis commando training centre.

Our assignment was to one of the largest border crossings called Pertuse. This was hugely popular with migrant workers, mostly North African, returning to France via Spain from trips back to their homelands. We were there to support the border police. Within minutes of arriving, the Legion began to enforce its no-nonsense approach. If any travel documents were out of date or not in order, the traveller did not enter France, no matter what their excuses. If anyone started to scream at the police or refuse to accept instructions, they immediately found themselves confronted by a pair of armed Legionnaires. Far from being grateful for the support, the border police were aghast at how their quiet, nod-and-wink existence was being wrecked. Whereas some police might have been willing to turn a blind eye when visas had expired by a few days or even weeks, the Legion made no exceptions – particularly when it came to Moroccan, Algerian or Tunisian nationals who had a conflict dating back more than a century with the *Kepi Blanc*.

We also took no chances with material being brought across the border and demanded that anything even remotely suspicious be opened, examined and, if necessary, verified. Pleas, explanations and curses from harassed travellers meant nothing; as far as we were concerned safety was paramount and nothing was left to chance. Sometimes I wonder just how many tragedies could have been avoided if others had adopted this approach to security at border crossings, airports, ferry terminals and train stations. And don't forget that this was fully six years before the 11 September attacks in the US, though France had already suffered terrorist attacks at the hands of Islamic extremists.

Eventually, the border police began to plead with us to take a more relaxed approach, not least because their small detention blocks were already full to capacity after just a few days of our operation. But there were times when the police realised just what an asset the Legion was. One day, a vehicle approached the border checkpoint at speed. This made the armed Legionnaires on duty take notice and, when the vehicle failed to slow down, almost as one we began to swing our FAMAS rifles into a ready position.

The van swerved across the front of the border stop and slewed into the inspection lane reserved for trucks and goods vehicles. The van shuddered to a halt in a squeal of tyres and burning rubber. Within seconds of it stopping, there were four Legionnaires surrounding the car, myself included, with our rifles trained on the driver. I still remember my finger pressing the trigger, waiting to apply that extra fraction of pressure for a controlled burst of fire at the first sign of a threat from the driver.

But, slowly, we realised that the driver was an old man – accompanied by his wife – who had probably enjoyed too much Rioja in the Spanish town down the road and thought he'd avoid any delays at the border by swerving through the goods zone. I can still remember the look on the old guy's face as it suddenly dawned on him that the Legionnaires pointing rifles at his head weren't joking. As we realised that he didn't pose a threat, we relaxed our aim, but we didn't stand down until he was taken out of the car and questioned by the border police. To this day, I'm sure the incident must have a bowel-loosening impact on the old codger.

As Legion duties go, Pertuse was pretty good. There were a couple of places in the near vicinity to get a few beers after duty and we escaped both the grim routine of barrack life and the back-breaking manual work that usually accompanied REG assignments. But, alas, it was not to last. I think the Legion truly frightened the border police and created no end of bureaucratic hassles. Within a short time, Paris began to reassure itself that a beefed-up border police could handle any concerns and the Legion were no longer necessary. But, for those few weeks high in the Pyrenees, we sent a strong message to any insurgents or terrorists trying to get into France: they'd have to come through the Legion first.

6

WITH NATO IN BOSNIA

My second tour in Bosnia was substantially different in tone to the first because this time the Legion was operating under a NATO rather than a UN mandate. So we swapped our blue UN berets for our traditional green Legion berets and, for the first time, our Kepis began to be worn. But the crucial difference was that NATO was prepared to use force whereas the UN had seemed almost determined to avoid its use.

That was fine by me because, like a lot of my Legion mates, we'd seen what was happening on the ground and were more than willing to show some of the militias that there was a big difference between 'ethnically cleansing' unarmed, helpless civilians and taking on trained Legionnaires. To be honest, there were more than a few of us almost praying for a showdown. I'd been sickened at what I'd seen being done to civilians, women and children. I attached tremendous value to the norms of military behaviour and I believed that being a soldier was supposed to be a noble profession rather than something shameful, which a lot of these Balkan militias had descended to.

The horror of what was going on in the Balkans was now dominating world headlines. There was a huge media presence throughout the parts of Bosnia and Croatia worst affected by the fighting. The siege of Sarajevo had aroused enormous sympathy throughout the world and now even special postcards were being printed to remind people around the world of what was happening in the city. Yet, in the full glare of the world media, the militias seemed determined to resort to any savagery necessary to achieve their goals before a general peace was agreed, or rather enforced.

Our second tour saw us based at Ranjclovak, an old Yugoslav Air Force base a lot further out of Sarajevo than our former complex at Skandarija. The major benefit was that it was significantly less exposed than Skandarija and the risk of attacks on our perimeter was significantly lower. By this stage, January 1996, the various warring Yugoslav factions had sat around a table at Dayton, Ohio, in the United States and had finally been pressed into sorting out their differences. But while the Croat, Serb and Bosnian leaders had agreed to a general peace plan and the rough outline of new national borders, the problem of dealing with the militias remained.

Some of them I judged as no more than gangs of criminals who acted as death squads. They were fighting for what they claimed were 'national' interests but were still resorting to everything from smuggling to extortion and prostitution. What made them exceptionally dangerous was the fact that they were now hardened to war and wouldn't hesitate to kill if they thought it necessary. Worse still, they were extremely well armed and had everything from AK-47 assault rifles to Russian-made Dragunov sniper rifles and rocket-propelled grenade (RPG) launchers. Some militia groups even had access to armoured vehicles from the now-defunct Yugoslav Federal Army. The only thing they lacked was air power – a factor that would prove crucial in the near future.

Making the whole situation even more complex was the fact that, appalled at what had been done to innocent, unarmed Muslim families in Bosnia, mujahedin-style volunteers were now pouring into the former Yugoslavia from all over the Middle East, including Iran. They wanted retribution, and they wanted to support their co-religionists in what threatened to escalate into an all-out 'Jihad' or holy war. It quickly became apparent that it was our job to stop them and to ensure the peace process was given a chance to work. But the arrival of well-trained and motivated Muslim military groups aroused very strong local sentiments. Memories faded very slowly in the Balkans. I often thought there were shades of the Irish mindset in the Balkan conflict because people talked about the catastrophic Serb defeat at Pristina to the Ottoman Turks in 1389 as if it were a key element of current politics and was precisely the reason why there was a Muslim population in Europe. Anytime I felt like dismissing the argument as insane I remembered how, back home, we still talked about the Battle of Boyne in 1690 and its relevance to modern Irish politics.

But now, for the first time since the civil war erupted, the various militias found that they would be taking on a serious armed force if they attempted to attack civilians or undermine the peace accord. Never before had the Croat or Serb militias faced the threat of heavy air-strikes or trained infantry assaults. We quickly learned that it was a threat they took very seriously and were ultimately reluctant to confront.

As an explosives expert, I found myself in constant demand – from dealing with a hand-grenade thrown on to a school playground which had failed to explode through to dealing with cluster munitions dropped by a withdrawing militia faction. It was exhausting work. The sheer scale of the munitions now being seized, discovered or simply abandoned was nothing short of staggering. NATO was stretched to the limit in trying to dispose of the ordnance. As a member of one of the disposal teams, I knew how important it was to stay focused and avoid tiredness. One mistake through exhaustion might not only cost me my life but could also kill my entire team. But it was also crucial to dispose of this ordnance quickly and efficiently so that it could never again be used to destabilise the Balkans.

It was soul-destroying work. And it also delivered one of the most eerie moments in my entire Legion career. In our section there was a young French Legionnaire by the name of Bernard. One night he told a few other Legionnaires that his grandmother had been adept at the 'dark arts' of palmistry and tarot readings. He reckoned that he had inherited some of her skills so, needless to say, we challenged him to put the talents to the test. I usually worked alongside two other explosives experts, Legionnaires Houlsek and Fotyi.

Bernard took our palms and started to read our fortunes.

When he got to me he abruptly dropped my hand and walked away, pretending to be upset. The other Legionnaires thought it was absolutely hilarious and soon the word spread that I was one of three people fated for horrific future problems. Unfortunately, over the next few weeks both Houlsek and Fotyi were injured on duty and I couldn't help but wonder what was coming my way.

Houlsek was almost killed when he was working on a clearance detail near a high-span bridge. Suddenly the structure came under sniper fire and Houlsek, seeking cover and also trying to spot the sniper, climbed along one girder of the bridge. He apparently reached up and grabbed an overhead sign to pull himself up on to the girder only to discover that the sign wasn't bolted down properly. It came away in his hand and he fell backwards, plummeting down on to a girder below. Houlsek was unconscious when other Legionnaires, including myself, got to him. He was bleeding from the ears and eyes and his elbow bone stuck out through his skin. It was a sickening sight. Most of the squad thought he was going to die but, incredibly, he managed to recover after several weeks in hospital.

Fotyi got injured on a mine-sweeping mission. He was checking a field for anti-personnel mines when he accidentally triggered one of the booby-trapped explosives. In the Balkans, this was almost always a death sentence. The Serbs and Croats had become so adept with mines that usually a small anti-personnel device was triggered into a whole batch of secondary explosives. Igniting the booby-trapped device usually meant that up to eight other mines scattered around the area were triggered – and your friends would need a mop and bucket to recover your remains. But when Fotyi triggered the booby trap the tension on the wire pulled the detonator right out of the sardine can in which the high explosives had been packed. The Legionnaire took the force of the detonator rather than the entire explosive device. He was injured, but had cheated death.

None of this made me feel any better. Legionnaires, probably because of the history of the force, did tend to indulge in superstition and by now I felt I was a marked man. I began to surrender myself to a dangerous cocktail of fatalism, frustration and anger. Fuck it, I thought, if I'm going to go, I'm not going to spend the rest of my time worrying about it. But, almost unknown to myself, because I believed that something bad was going to happen I began to relax my own precautions while out in the field.

The day we got a call about a hand-grenade being thrown into the playground of a local school, I'd had enough. 'What kind of scumbag throws a grenade at schoolchildren?' I asked my Sergeant. 'Fucking barbarians!' I said, disgusted. 'They're fighting a war against mothers and babies.' When we arrived at the school, the rest of the team began the laborious task of putting on their protective gear. But I was pissed off and I was losing my faith in humanity fast. Worse still, I had no-one to take out my frustration on. I walked right over to the hand-grenade and, taking a lump of plastic explosive, I began to shape a cover charge to place over the grenade so we could explode it safely. I quickly moulded the charge into a shape like a bridge and then placed this over the hand-grenade. By now the entire team were staring at me in horror from a safe distance. I was breaking every single rule in the 6th REG safety manual. I hadn't put on my protective gear, I hadn't stopped to analyse the scene for hidden dangers and I'd allowed myself to be stranded too close to the munition that I was trying to explode.

Because I'd rushed the shaping of the charge, it wasn't properly balanced and immediately slipped to one side, rolling the grenade over with it. I knelt there in my regular combat fatigues, just inches from the grenade, and watched motionless as it gently rolled to a halt. But it still didn't explode. Almost at once I realised how incredibly stupid I'd been and my Sergeant-Chef took me aside immediately afterwards and warned: 'No more shit like that, Paddy. Next time you do it by the book with all the gear on.'

I had walked away unscratched from my closest call with death so far. But I did pay a price. The incident became the focus of the one and only nightmare I've ever had from my military experiences. Every so often if I'm tired or under stress, I'll relive that day in the playground and wake up drenched in sweat as, in my dream, the grenade explodes with a roar the very instant it stops rolling. It's so real that when I wake up I am convinced my ears are ringing from the force of the explosion.

A fortnight later, we were working on another mine-field in the countryside outside Sarajevo, but this time using special inflatable gear called 'M-Ms' (Matra Mines or Mine-Mats), which were designed to disperse the weight of our footsteps and thereby avoid the triggering mechanism for anti-personnel mines. The equipment was apparently state-of-the-art, but I felt a little like a guinea pig. I knew they had been used in the Gulf War just five years before but I also knew that in Kuwait and Iraq they were used on flat, desert surfaces. Here in Bosnia, the mine-fields were laid in rutted, uneven and stone-pocked fields, substantially different from desert plains. It was very awkward moving in the gear which, because of its function, was filled with eight air-pockets. It was like trying to walk around in a pair of giant inflatable slippers.

In either hand I held two special testing poles designed to probe the ground for any indication of mines and to erect special warning signs afterwards. Each had a sharp spike at the end and, trust my luck, didn't I accidentally puncture one of my inflatable 'slippers' with a spike. So I was now standing in the middle of a minefield with one leg boasting a weight-dispersing device and my other leg resting, like normal, on my boot. All I could do was raise my eyes to Heaven and silently curse myself. My partner kept working and I had no option but to walk on with him. A few strides later my blood froze with the distinctive grating sound of the probe of an anti-personnel mine scraping along the bottom of my boot.

The mine had been planted in a small hollow which I hadn't spotted. The detonator had scraped off the underside of my boot and I was now mincemeat. I stood there motion-less, waiting for the percussion of the exploding mine to blow me off my feet. But there was nothing but silence – the only sound was the manic beating of my own heart. It felt as if my heart was caught on my Adam's apple, but I forced myself to remain calm. I slowly raised my hand up to indicate a mine contact, and the rest of the team froze in position. I can't remember how long I stayed motionless but it seemed like forever. Then, slowly and in agonising suspense, I lifted my leg away from the contact. And still nothing happened. By now, the sweat was pouring from my forehead in a torrent. As I stepped away from the anti-personnel mine I felt wasted.

We later conducted a careful examination of the mine and found that, just as my worst nightmare had imagined, it was part of an inter-connecting field of mines. If it had exploded, it would have triggered a wave of other mines within a fifty-metre radius. I would have been eviscerated. But, incredibly, the mine I stepped on had obviously been planted some time before and the mechanism, exposed to the harsh Balkan winter, had rusted. When I triggered the detonation mechanism with my boot, the firing pin had been kept in place by the rust and it blocked the detonating mechanism.

Later that night I was back at the base and feeling as if I'd enjoyed divine intervention. Legionnaire Bernard walked up to me and, winking knowingly, said: 'You can relax now – it is over.' Without saying another word, he walked away. Houlsek and Fotyi both made full recoveries and I learned the lasting lesson that you walk through mine-fields very, very gently. It's hard to accept it, but I owe my life to a stubborn piece of rust.

Finally, after several weeks of handling munitions, it was with relief that we were informed that our company would be supporting a special NATO operation in the Drushina valley. Together with Caporal-Chef Oberle, I readied my munitions kit, and prepared my full infantry gear, including my FAMAS and a combat load of rifle ammunition. Gradually, the word filtered down to our section that it was a hugely sensitive mission with potential implications for the success of the entire Dayton peace accord.

Apparently, NATO intelligence sources had located a mujahedin 'cell' in an old ski lodge in a remote part of the Drushina Valley. This cell was being trained by Iranian insurgents and, NATO feared, was preparing a campaign of retribution against the militias that had been massacring Muslim civilians. NATO sources had described the Drushina cell as 'a training school for mujahedin or holy warriors'. Now, the raid on the cell would be led by crack NATO units including Britain's SAS and France's unfortunately named special action commando unit, CRAP (Commando Recherche et Appui dans La Profondeur). The CRAP units were drawn from the Legion and were renowned for being the elite within the elite. They've now been renamed as GCP-SPECIALE and are at the front of France's anti-terrorist and special operations elite.

The SAS and Legion commandos would form the strike team that would seize the ski lodge and the surrounding hamlet; it would then be our job to sweep the whole valley, secure the area and protect the strike team from any counterattack. The operation was one of the most dramatic I'd ever seen. There must have been fifty helicopters, from transport SuperPumas to heavily-armed gunships. We were airlifted into the Drushina valley and, while the strike team dealt with the ski lodge, we were to secure the area and search each property for any sign of a threat.

The giant air armada swept into the valley. It must have been an awe-inspiring sight for anyone on the ground. Together with my 6th REG squad, I jumped out of the chopper – receiving a bad back injury – and began the laborious task of a combat march into the secured area before checking on each and every house, garage and barn for signs of armed resistance. We had to leave our full kit behind under armed guard. Not long after we began our work, the word spread that the strike team had captured three Iranian nationals in the lodge. But, even more seriously, there was apparently a large force of around two hundred mujahedin in the wooded hills directly above us. They had witnessed the strike team's arrival and were now threatening to mount on all-out assault if their three Iranian friends were not released immediately.

But that was easier to threaten than to deliver, given the huge NATO force now tightening its grip on Drushina. The strike team had not only secured the three Iranian nationals but had uncovered a nightmarish cache of weaponry. This ranged from wiring devices for car bombs through to booby-trapped dolls and toys. An onslaught was clearly being prepared. The cache was later described to us as an Aladdin's Cave for terrorists. Given the delicate stage that the Dayton Accord was now at, if such a cache had been used it would almost immediately have undermined the ceasefire and provoked some of the more vicious militia groups to hit back in retaliation. For once,

NATO intelligence had proved unerringly accurate – and hugely important.

Now, the objective for NATO was to get the Iranians out of the valley safely so that they could be questioned properly by the special gendarmes that always accompany the Legion on its foreign missions. The French commanders realised the full importance of their catch and wanted to debrief the Iranians in detail about what other armed training camps were being developed in Bosnia. But they also knew they had to get them out of Drushina fast.

Having confirmed that the area was secure, my unit prepared to provide a special security cordon for the strike team as they brought the three captives to a location outside the hamlet for airlifting back to a NATO base. The huge advantage of using choppers to exit the valley was that we totally avoided any risk of a roadside ambush. While there was a risk of ground-to-air missiles, the helicopter gunships that hunted hungrily around the valley for any sign of a threat made such an attack unlikely. For the attacker, any missile launch in the face of such gunship power would have been virtually suicidal.

Our Legion unit lined the streets and, as the armada of helicopters began to arrive to bring the troops back to base, we maintained a careful vigil, with every Legionnaire having his FAMAS at the ready. I have to say that the security cordon looked impressive and, if the mujahedin had launched an attack, they would undoubtedly have received a nasty surprise. To this day, the image of the Iranians marching through the Legion lines towards the waiting choppers is vivid in my memory.

The final strike team to withdraw was the Legion's commando unit and, as they walked to the waiting choppers with their three Iranian guests, they looked a pretty tough bunch of bastards. I knew they were based at Calvi on Corsica and that their training regime rivalled anything the SAS or German GSG9 enforced. But what I didn't know was that one of my future great friends was part of the CRAP team that day, though while we marched past each other in the looming Balkan darkness we didn't actually meet. It was only ten years later than I finally met Denis B. and discovered that we'd marched within metres of each other in the Drushina valley that day.

Drushina proved to me what a well-led, efficient and motivated force could achieve in a situation like Bosnia. But, if the militias thought they were facing anything less, then there would have been tragic consequences.

Reflecting back on my first mission in Bosnia I realised that in contrast to this one it had been handled badly. The earlier presence of the troops under UN control had achieved little to calm the situation, as the tragedy of Srebrenica graphically proved. The Legion is a tough combat unit, but as part of the UN mission to Bosnia we had effectively been wasted. We had little flexibility of action and almost every move was double- and triple-checked by UN command. At times I felt that we were like a toothless Doberman – lots of noise but very little bite. During several UN convoy escort missions, we arrived at our final destination with virtually empty trucks because at every Serb,

Bosnian and Croat checkpoint en route the local militias had unloaded their share of our cargo. We had to stand by and watch helplessly. I hated what I saw. We also knew about the ethnic cleansing and massacres of civilians in the hills. But it was only now, with the benefit of hindsight, that I could get a perspective on so much of what was happening around me during my first tour. As so often happens in brutal civil wars, the truth takes some time to emerge either through the media or through international investigations. We were only now learning that had really happened in those Bosnia hills to unarmed men, women and even children – and many of us Legionnaires were losing patience with the 'softly, softly' approach.

Throughout the entire Balkans civil war, Western soldiers paid a price in blood. A significant number of Legionnaires were maimed or badly wounded, most from sniper fire or minefields. Even though I shudder to think about it now, the situation was so serious in early 1996, just as we commenced our second deployment to Bosnia, that Legionnaires within my company made an impromptu agreement that any soldier who was crippled or horrifically wounded would be given the opportunity, by his comrades, of ending his suffering. We all took this to mean that if one of us was maimed, another Legionnaire would have to provide the wounded trooper with access to a pistol. It may sound horrific, but we sincerely believed that we all should have the choice about how to cope with our injuries, particularly if they were so bad that you'd be left with little or no quality of life.

Mercifully, this never arose in Bosnia. Fotyi escaped

serious injury and made a full recovery after spending several weeks in hospital. Houlcek likewise recovered, which still astounds me because I was the first to reach him after he fell from the bridge girder and, given the terrible injuries he sustained, I didn't think he would pull through.

But above all, Bosnia was an absolute sewer for moral sensibilities. It was a conflict where all the rules were thrown out the window and even the most ordinary of situations had the potential to create tragedy. I learned this to graphic effect on my first tour there when we were providing an armed escort for three dump trucks through Sarajevo. We were to bring the trucks to a designated dump for all the domestic rubbish from the city as well as all the debris and rubble from the war. It was as routine a duty as you could imagine – in fact, some of us wondered at the sanity of providing escorts for what were little more than over-sized bin lorries.

Incredibly, as we drove into this sprawling dump on the outskirts of the city, I realised that there were people living here. In fact, hundreds of families called this place home and they survived by living off the rubbish. The stench was unbelievable and it was as close to a picture of hell as you're likely to get in modern Europe. The instant we brought our APC to a halt, it was swamped by people – mothers, young teenagers, children, all begging and crying for food.

On every mission to that place I know that the Legionnaires emptied their ration packs to the youngsters and, in a mad scramble, the kids would devour whatever food we gave them right there and then. The packs would be ripped open and the food shoved straight into their mouths. Those who got the food would walk away satisfied; those who missed out would howl for something – anything – to eat. It was the closest thing I had ever seen to humans being reduced to the state of virtual animals.

In that situation there was no such thing as innocent childhood either. One day, I was driving the APC as part of the escort detail with my commander sitting beside me and three armed troopers in the back. It was such a routine mission that the troopers had battened down all the hatches, closed the firing ports and were snoring loudly in the back. However, on the first trip down a hill by the entrance to the dump I thought I heard the 'zing' of a bullet ricocheting off the APC's armour plating. But over the noise of the diesel engine I wasn't certain. I mentioned it to my commander and he said he hadn't heard anything. On our next trip, at the same spot I heard the same thing but, again, my commander dismissed it as nothing. By our third run, I was totally focused on the sound and, sure enough, as we passed the spot I heard another 'zing'. This time, I slammed on the brakes and brought the APC to a shuddering halt, waking the dozing troopers in the back as they were tossed violently against the armoured hull.

I grabbed my FAMAS and jumped out of the driver's seat, scanning the area for a target. Immediately off the road was a small, ruined house that obviously was some family's home. I carefully moved behind the building to check for signs of a sniper and spotted a small boy, standing right in front of the rubble of the shack, holding a rifle. I carefully looked through the sights of my FAMAS and realised that the boy, who was grimy and dressed in filthy clothes, could have been no more than eleven or twelve years of age.

The rifle he was holding was a small-calibre .22 hunting rifle, which explained why the sound of the ricochet was so slight. But he was shooting *at us*. I carefully took aim at his upper torso. His face seemed to freeze in shock; he simply stood there, mouth open, almost waiting for what was going to happen next. My commander had also jumped out of the APC and ran up to me whispering urgently: '*Merde*, it's only a kid, it's only a kid.'

But my finger was on the trigger and just a fraction more weight would send a high-velocity round right through the junior sniper. Slowly, I eased off the pressure on the trigger and exhaled long and hard. One second I was ready to kill to defend my colleagues and the next, mercifully, reality dawned and I realised that I didn't want a child's life on my conscience. You little bastard, I thought to myself, we're here to help your city and you're shooting at us – this isn't a fucking game. I slowly lowered the FAMAS and the kid suddenly started shaking with fright. I just walked away. I couldn't even bring myself to take his rifle from him. But that was Bosnia – even the most routine, menial jobs could cost you or someone else their life. You only had seconds, sometimes even micro-seconds, to make a decision about what you were going to do. But then you had the rest of your life to live with that decision and it remains every soldier's worst nightmare to kill a child, even one pointing a rifle. Which is why Bosnia was such a mentally draining and emotionally exhausting place to serve on armed missions. There was no black-and-white, merely thousands of different shades of grey, and the terrible memory of standing over so many flag-draped coffins and learning to ignore your emotions.

But it was hard to ignore events like Srebrenica. The UN safe haven proved little more than a graveyard for virtually an entire generation of Bosnian men and boys in July 1995. The troops that were guarding them – a Dutch detachment, under UN control, which was largely composed of conscripts – stood aside and allowed Bosnian Serbs under General Radko Mladic to 'remove' almost 20,000 Muslims. More than 8,000 Bosnian men and boys were later murdered in the mountains and buried in mass graves. The Dutch troops were outnumbered and out-gunned. But they completely failed to stand their ground – and just stood by and watched as the Bosnian Serb militia separated the men and boys from the women. Some Bosnian Muslims have claimed that the Dutch actually facilitated the separation process. Later, the Dutch unit, operating under UNPROFOR (UN Protection Force), fled to the safety of Zagreb. One Israeli study compared the Dutch action at Srebrenica to the massacres at the Sabra and Shatila camps in the Lebanon in the early 1980s.

The result was the worst massacre in Europe since World War II. The hundreds of women and young girls who were separated from their fathers, brothers and husbands were subjected to brutal rapes and sexual assaults. I wondered: had I managed to grab my rifle that night in the police station with Karadic might things have been any different? And I also wondered could the Legion have made a difference at Srebrenica?

Legion units in Bosnia had been placed on high alert but, at the time, we didn't know why. It was only afterwards we realised that had the Dutch force resisted the Bosnian Serbs we had been earmarked as a combat unit to tackle the Serbs in Srebrenica and prevent atrocities. The story I heard afterwards in the Legion canteen was that there was no air transport available to ferry us from the mountains to Srebrenica. Similarly, no other Western countries were able to supply air cover for our operation. And so eight thousand Bosnian men and boys were marched away to their deaths – and the West never fired a shot.

Would the Legion have made a difference? Even if we had been deployed, we wouldn't have been able to match the Bosnian Serb artillery and would have been entirely dependent on air support. We would also have been substantially outnumbered. But the Legion's history was made in defying the odds, and whatever those odds the eight thousand men and boys would at least have had some chance. Bosnia convinced me that UN and NATO peacekeeping can be a bit of a joke. Maybe it suits some armies, but the Legion is a combat outfit. The Legion simply should not be put into an environment where its training can make a difference and then have Legionnaires stand idly by and not be allowed to put their skills to use.

7

QUITTING MY
BELOVED LEGION

I didn't leave *La Legion*. As far as I am concerned, *La Legion* left me. But the scars over that parting still hurt even now, almost ten years later.

To this day, I'm haunted by precisely how my dream life turned into a nightmare and I often wonder could things have been different. In 1995, I loved the Legion and wanted to spend the rest of my adult life wearing the Kepi. I had earned my Caporal's stripes and I knew making it to Sergeant level was only a matter of time. I knew I was a good Legionnaire – maybe a bit cocky, a little outspoken, but cut from the cloth that had made the Legion what it was.

I tried to live my life along the lines of honour and courage that the Legion demanded. In return, all I wanted was to be able to believe in what I was doing and trust the men giving me the orders. And then came Bosnia. The frustration of standing by helplessly while horrors were committed all around you, knowing you had the power to make a difference but not being allowed, for political or diplomatic reasons, to use that power. At the end of my first tour in Bosnia, the frustration was like a volcano inside me. I had to let off a little steam and, unfortunately, Lt Guyot chose that precise moment to come back into my life. But he was now a Captain and clearly he hadn't forgotten the training incident from Lardeche, and though I hadn't been the one to lose the rifle, I seemed to be 'guilty by association'.

After my first tour in Sarajevo, I was ordered with the rest of the 6th REG back to Avignon. We were rotating back to France via Zagreb where part of the city had been set aside for UN use. As a Caporal, I was responsible for a section and, on arriving in the Croatian city, I warned the Legionnaires about not going into town. It was our last night in the Balkans and the senior officers didn't want any problems. We would all get leave when we got back to Avignon. So, like all the other duty Caporals, I warned the men to save their socialising until then. They could relax inside the UN compound, but no-one was to go into Zagreb.

Then, after dismissing the men for the night, I immediately broke my own directions. I found Ascot and Tabone and went looking for somewhere to have a few drinks. We couldn't go off the base and discovered that the only place to get some beers was in a BurgerKing franchise that had been opened by two enterprising young German guys in a converted articulated lorry. They made a small fortune over those few weeks. All the other bars on the base had shut at 7.00pm.

We got stuck into the beer on offer and, a few hours later, we finally got ourselves invited into a Swedish officers' club on the base. The club was already full of French marines; I knew a lot of the lads because I had worked side-by-side with them in Sarajevo. One of the guys was the marine who had ordered me to do compound perimeter checks during the militia assault in Scandarija – and he owned me a big favour. Because I was the mine specialist, the marines obviously reckoned they also owed me a few drinks so they started buying whiskey for myself, Ascot and Tabone.

While everyone else seemed to be unwinding and relaxing, I felt as if a huge pressure cooker was about to explode inside me. I was so pissed off with everything – from the horrors of Sarajevo to the futility of the UN mission and the Legion officers who cowered before their regular French Army superiors. What had started as a few beers before heading home was now a ferocious, no-holds-barred session. At one point, a Swedish military police (MP) unit came through the club, and they made no secret of the fact they didn't like the way we were partying. Those Swedish MPs were built like wrestlers and had the old, German-style collar plaques to indicate their rank. They were blatantly trying to intimidate us, but neither the French marines nor the Foreign Legion are known for being easily intimidated.

Finally, we realised that another Legion friend, Hannier, was on the point of collapse so we decided to drag him back to the billet. Ascot and myself grabbed Hannier by the arms and started to carry him towards the Legion barracks. We hadn't even gone ten metres from the club when the Swedish MPs confronted us. They spoke in Swedish – but we knew they were giving us shit about the noise and our drunken condition. I had swapped my Legion top with a Swedish soldier and this clearly didn't help matters much. In my jumble of uniforms, I probably looked like a mongrel to these immaculately turned out MPs. Unfortunately for those Swedes, they didn't realise who they were dealing with and, almost in unison, Ascot and I dropped Hannier and waded into them.

There were four of them and within minutes we had torn through them. I knew I was losing control and was taking out all my frustration, anger and rage on the Swedes. We hammered them – but almost immediately, reinforcements began to arrive. Soon Ascot and I were facing about eighteen American, Swedish, French and Finnish MPs. But we were fighting mad now and all the combat training had turned us into dangerous brawlers. Ascot was a brute at any time, and I had completely lost the plot. I was determined to take every bit of anger I had out on the MPs. We had taken about eight MPs out of the brawl before their numbers finally told.

Wisely, the MPs barricaded the door to the Swedish club because the French Marines and Legionnaires were desperately trying to get out to help us in the brawl. God knows what would have happened if they had managed to do so – it probably would have been a bloodbath. Finally, the MPs managed to get us on to the ground and, after wreaking some savage revenge on us with their fists and boots, they handcuffed us. To make matters worse, my Foreign Legion Lieutenant, who had been called to the scene to see what was going on, was arrested, forced to the ground and hand-cuffed as well. It took fifteen minutes of discussions before the MPs finally accepted that he had nothing whatsoever to do with the incident.

I didn't help matters by screaming obscenities at the MPs for daring to arrest one of my Legion officers. To them, he was simply another Legionnaire. One Swedish MP, who had seen three of his colleagues hospitalised by our attack, simply grabbed me by the hair and dragged my face along the ground towards a police car. Half the skin of my face was left on the pavement. That wasn't enough for them and the MPs checked my handcuffs, making sure they were so tight that I began to lose all feeling below my shoulders. The last thing I said to the Swedish giant who threw me into the back of the police car was: 'Don't be near me when these cuffs come off because I'll kill you.'

By now, the volcano of rage inside me had finally blown. I went mad in the police car and tried to kick out the windows. Eventually two MPs had to sit on me for the short journey to the detention block. I was thrown on the floor of the reception area where a US Marine Corps MP was to process our details such as name, rank and units. Ascot was calm by now and started chatting to the MP about the Corps and the fact that he, too, was American. But I was far from finished. What the fuck, I thought, let's have another go. So I jumped up, hands still cuffed behind my back, threw myself over the reception desk and proceeded to demolish filing cabinets, desks and typewriters with my flailing legs. It took three MPs to restrain me – and I was locked in a bare cell for the night. I still laugh when I remember the look on the Marine Corps MP's face as I was led away. He looked at

Ascot and rolled his eyes to heaven and said: 'Foreign Legion – holy shit!'

By now, the Legion command had heard all the details of what had happened and regarded the incident as a major embarrassment. In particular, the officer in charge of the Legion's rotation back through Zagreb was out for blood. Typical of my luck, this was Captain Guyot. I was returned to the Legion contingent in handcuffs, and my face looked like it had been run over by a lawnmower. There were blood stains all over my combats, but two Legionnaire friends had managed to smuggle a spare top to me in place of the ragged Swedish top I had been wearing. Unfortunately, the new Legion top was too small for me, so I truly looked disastrous when I rejoined the 6th REG. A leper would have received a warmer reception from the REG officers.

To the ordinary Legionnaires, Ascot and myself were heroes and the brawl was fully in keeping with long-standing Legion tradition. But Captain Guyot was furious and as we walked towards the plane to take us home he privately promised me I was going to prison. 'Forget about leave, when we get back to Avignon you're going straight into a cell. You're finished. There is a stack of charges against you,' he warned.

To be honest, I was terrified that he was right. I knew that a few of the MPs had been hurt and that the Swedish, Finnish and American authorities were screaming for blood over our actions. On the way from the airport back to our French base, the thought of going AWOL crossed my mind. But, bruised and battered as I was, the Legion was still my life.

Over the years, the fighting quality of its soldiers was some-thing that the Legion traditionally valued – and, luckily for me, our Colonel back in Avignon was very much 'old' Legion. At the muster parade on our return, I stayed in the rear ranks hoping my bloodied and bruised face wouldn't be noticed. There was beer and wine for all the other Legion-naires, but, not surprisingly, there was nothing for me. I knew that the second everyone else was dismissed, I would be ordered into the Colonel's office for punishment.

I was in full No. 1 dress uniform and the white of my Kepi stood in stark contrast to the black and blue colour of my face. The whole side of my face was covered in bloodied scabs and every inch of my body hurt. The Colonel was sift-ing through a disciplinary file that looked a good six inches thick. He read off the charges facing me, and I simply replied: 'Guilty as charged, Sir.' But, to my amazement and to Captain Guyot's horror, the Colonel remarked on my exemplary tour records from Bosnia and Cambodia. In particular, he men-tioned the fact that I had been in very stressful situations and had performed well for the Legion. It helped enormously that I also had a formal letter of commendation for my service in Sarajevo. I couldn't believe my ears when the Colonel imposed a six-month suspended sentence and told me I was free to go. Incredibly, I also got to go on leave with the rest of the 6th REG. I knew that Captain Guyot was fuming, and I took a malicious pleasure in it. But I also said a silent prayer that our paths would never cross again.

Unfortunately, twelve months later that is precisely what happened. The brawl in Zagreb at the end of my first tour in

Bosnia had started my disillusionment with the way the Legion was changing and, despite a second Bosnia tour, I felt that the old Legion I knew and loved was gradually being transformed to fit the needs of the regular French Army. In the space of two years, there was a haemhorrage of Caporals and Sergeants – the very backbone of every military outfit. Some of the guys that I ranked as amongst the best in the Legion were leaving, including several of the Sergeants who had trained me. Slowly but surely, the old Legion traditions began to be replaced by new systems. Even the old pay system, where you reported for your monthly salary and scooped the cash into your Kepi, was abolished in favour of automated monthly lodgements into your bank account.

And the new recruits began to change as well. There was an increasing number of East Europeans coming to Aubagne and Castel, and, while some were very good soldiers, they brought with them a suspicion that the Legion was now simply a route to secure a French passport. It also began to emerge that Legionnaires would have to undergo various academic tests and assessments, with the classroom replacing the hard school of The Farm and Castel.

But it was the junior officers that gave me the greatest concern. They were opting for the Legion from the regular French Army because it would further their careers and promotion prospects. They didn't give a damn about the Legion or its traditions and, whereas once only the very best from St Cyr were allowed rotate to the Legion, now the option was opened up to all officer candidates. It was too much to bear.

The 'old' Legion that I had joined was being transformed into the 'new' Legion to better fit with France's military requirements. For generations, *La Legion* had provided a hardened professional corps of soldiers for the Republic and, because the French Army relied heavily on conscripts, such a unit was absolutely vital. But the French Army was now moving to adopt the British model of an all-volunteer, professional army. The purpose of *La Legion* was deemed by the Elysée politicians to have changed.

There has always been an uneasy relationship between the Paris politicians and their army of overseas Legionnaires. The Legion felt itself betrayed in south-east Asia by the very politicians who had asked them to fight an impossible war but denied them the resources to fight it properly. The Legion regards its soldiers as a family and no member of that family is ever left behind, but hundreds were abandoned in Vietnam as French forces pulled out of Hanoi after the Dien Bien Phu defeat. French military personnel sent out into the mountains and deltas to help arm tribes against the Communist Vietminh were simply abandoned. Their ammunition supplies were cut, their radio links were switched off and they were left to fend for themselves as Communist hunter-killer teams closed in. One horrific story tells of a pilot flying in Thailand along the Vietnam border two years after Dien Bien Phu and, over his radio, hearing the curses of a French soldier in the jungle below who simply wanted ammunition so as to be able to die like a man.

Algeria also involved a huge sense of betrayal. Algeria was not seen as a colony – it was deemed to be as much a part of France as the Dordogne or Poitou. But it was also the Legion's fabled home with all the Legion flags and memorials stored at Sidi-Bel-Abbès. After the order to quit North Africa, part of one Legion Parachute regiment rebelled against the Paris government of Charles de Gaulle. The rebellion was put down, the regiment involved was disbanded and the Legion was substantially trimmed in size. But the sense of mistrust, on both sides, was never really forgotten. The Legion had spilled too much blood over the years to ever fully trust politicians. France resisted calls for Algerian independence and then, after the Legion had lost thousands of men fighting against a bloody insurgency campaign, had done a u-turn and granted Algeria its statehood. In Legion parlance, while the politicians dithered Legionnaires died.

But now even the commitment of the very backbone of the Legion – the NCOs – was changing. For more than a century, the Legion traditions, legends and *esprit de corps* were jealously guarded by its Caporals and Sergeants. I was shocked when, after one route march with the French Marines, we were effectively abandoned by our NCOs as we mustered outside the base. Partly because of its colonial origins, the Legion marches in a different stride and cadence to both the marines and French Army. The Legion marches at 88 paces per minute compared to the average march of most European armies, which is at anywhere from 120 to 140 paces. It is impossible for a Legion unit to march in tandem with a regular Army unit if it is to maintain its traditional marching cadence. Now we were ordered to march into the base alongside the marine units. What should have been a proud end to our day ended up a total shambles, with the Legion units looking disorganised and unprofessional alongside the Marines as we tried to march to their tempo.

Something like that would have been unthinkable in the old Legion. We were famed for our marching, and for the songs we sang as we marched. No matter what the manoeuvres or the exercise, a Legion march was something to behold. We had an image, an ethos to uphold. The Legion's unique rhythm was aided by a set of marching songs that proudly traced Legion history from North Africa to Indochina, and were zealously protected by Legion sergeants. Legion units practised to the point of exhaustion until they marched in perfect unison. So impressive was the sight of a Legion unit marching in perfect step back to barracks or on to a parade ground, proudly singing their famous songs, that everyone – whether Legion soldier or ordinary civilian – always stopped to watch and admire. On this occasion our NCOs had let the ordinary Legionnaires down in a despicable way. Superior rank or not, I told a few NCOs the hard truth later that night. But the reality was that, with all the changes and political uncertainty over the Legion's future, they just didn't care.

For generations, the Legion had maintained that it was the 'family' of every Legionnaire: no matter what the problem, the Legion looked after its own. That attitude had been carefully handed down from Sergeant-Chef to Sergeant and from Caporal to Legionnaire for more than a hundred years. Legionnaires were told that they might have to serve some time in military detention for their crime but, except for the gravest offences of murder or treachery, the Legion would still remain their home. That was a tradition that was in stark contrast to other military units where the threat of dishonourable discharge was always held over a soldier's head.

But this was no longer the case in the Legion. I learned that fact one day in 1996 when a dispute in the Avignon barracks between two Legionnaires – a Vietnamese called Hanto and a Romanian recruit – cost one of them his Legion career. I was in the barracks that day when we heard a shout and some commotion from the recreation room. I walked down to see what was wrong and was astounded to see the Romanian lying groaning on the bed, holding his abdomen. The little Vietnamese Legionnaire was sitting calmly on a chair in the middle of the room watching *Little House on the Prairie* on TV.

When we checked what was wrong with the Romanian, we realised that he'd been stabbed – with a precision that a surgeon would have struggled to match. Apparently, a dispute had erupted between the duo over which channel to watch on TV. When the Romanian, who towered over the little Vietnamese, tried forcibly to change the channel, Hanto whipped out a blade and stabbed him. Leaving the Romanian groaning on the bed, he cleaned and stored the blade, before sitting down to finish his TV show.

Luckily, the Romanian made a full recovery; I don't think Hanto wanted to seriously hurt him. In fact, the Vietnamese was such an incredible fighter I think he could have filleted the East European if the urge took him. Despite his size, he was absolutely fearless – one of the toughest men I've ever come across. He never shirked on a march and was a born soldier. At the time it struck me that if the North Vietnamese fielded an entire army of guys like Hanto, not even Caesar and his 9th Legion could have beaten them.

In the old Legion, Hanto would have served a nightmarish time in military detention – and, believe me, there was nothing pleasant about a Legion prison. He would then have been assigned to some particularly awful tour of duty, possibly in a God-forsaken corner of Africa, the Pacific or South America. If the Legion thought he enjoyed overseas postings, he would probably have been assigned to cleaning parade grounds or toilets for a few years. He would never have been considered for promotion. But he would have remained a Legionnaire.

However, as it turned out, this was the new Legion where an entirely different meaning was attributed to the word 'family'. Hanto served his period of military detention and was then unceremoniously thrown out of *La Legion*. It was a particularly savage blow for the little guy because, from what I could gather, the Legion was his entire life. He had nothing else – no family, no contacts in France and little hope of being able to go back to Vietnam. I was never friendly on a personal basis with him, but I was still gutted for him. Once he was thrown out of Avignon, we never heard another word about him. As far as I was concerned, it was another betrayal of the old Legion code of honour.

Yet I knew I had great prospects within the Legion. I was earmarked for Sergeant and, before I went to Bosnia for the second time, I told my superiors that I would re-enlist for a further five years. I knew I was respected by ordinary Legionnaires, and I was damn good at what I did. But I came back from Bosnia with doubts for the first time about spending the rest of my life in the Legion. I wasn't a politician and I wasn't a promotion-route soldier. I loved the Legion and all it stood for, but I couldn't stand by and watch as my beloved Legion was prostituted into something else. I arrived back in Avignon and was confronted with papers to re-enlist for the next five years.

By now, a lot of the changes in the Legion had begun to grate on me. But I was still prepared to put up with them because I had my own ambitions. I had for years been wanting to go on a security-dog training course with the Legion as I had always loved dogs. I'd made my feelings known to my superiors and, with my Sergeant's course also on the horizon, I reckoned I had a very good chance of getting approval for it. But, just days before I was due to confirm the re-enlistment papers, a friend from the Avignon base office contacted me and said that, unfortunately, no Legionnaires would be sent on the dog-training course for the foreseeable future. Bluntly, he admitted it could be years before the course became available for members of the Legion.

A short time later I met Captain Guyot and, when reminding me about my re-enlistment papers, he said that my place on the security-dog training course had been booked and was waiting for me. All I had to do was sign the final papers to re-enlist for another five years. Instantly, I guessed that I was being sweet-talked – and misled. And I also realised that my re-enlistment was now being used as a tool to ingratiate my superior with the Legion commanders. It was something I found impossible to forgive, particularly after the blood, sweat and tears I'd put into the Legion.

So I stalled on signing and the final straw came when it was confirmed that my new permanent Company Captain was going to be none other that Cpt Guyot. When he finally confronted me about my future plans, I told him honestly that I didn't think I was going to sign. Not surprisingly, he freaked out. By the time I left him and got back to my barracks, I was informed that I was being transferred out of the unit. It was like I had been physically slapped in the face. I had gone through hell with these guys in Cambodia and two tours in Bosnia – and now I was no longer needed? But I was told that, because I was not re-enlisting, I was being taken out of the combat unit and sent to the company command. I still had almost six months to serve and, perhaps naively, I thought they would leave me within my own combat unit.

Now, I was to be administrative fodder for the rest of my time with the Legion. I would either spend my time twiddling my thumbs in an office waiting for office duties or, more than likely, I would be assigned to a senior officer to act as his driver and part-time coffee-maker. For a full month, I sat in the administrative offices in Avignon and did absolutely nothing. I thought I would lose my mind. To an adrenalin junkie like me, it was a fate worse than death. And then I heard that our regiment was heading back up to northern France for some major manoeuvres with the French Army and Air Force.

I was given duties to assist with the REG's vehicles and I took it on myself to get a VAB armoured personnel carrier with a burned-out engine ready for pre-manoeuvre inspection. I worked on that wrecked armoured personnel carrier for weeks and was enormously proud of myself when it passed, with flying colours, a military inspection for inclusion in the manoeuvres. If I was going to leave the Legion, what better way than on manoeuvres in northern France, screaming around forests and mountains in a VAP that I'd helped repair?

And then Captain Guyot decided that I was needed by his side. I was going to be his personal driver – and, no doubt, responsible for sourcing the Captain's espresso or cappuchino. From being able to pay a final, proper farewell to the Legion, the five-week-long manoeuvres would be an exercise in humiliation for me. 'Fuck this,' I said to myself. And that was it. I walked out of the base and never went back. I had less than four months left on my Legion contract – but I simply didn't care. I hadn't left Ireland and walked through those gates at Aubagne to be some officer's servant. I hadn't gone through the brutality of Castel to make coffee for some St Cyr career politician. If I'd wanted to make coffees for a living I would have stayed in the hotel kitchens in Cavan.

So I left. I didn't even bother to empty my gear from my locker. I just walked into Avignon, booked into a hotel and went on the town. For the next few days I drowned myself in beer and *pastis*. A few of my friends tracked me down and joined me on my session. The lads even organised for my gear to be brought from the base to my hotel. A few days later, I went up to the bank to get my cash and close my account. The branch was literally right next to the front gate of the base. But no-one came after me. The Legion knew where I was and, it seemed to me, they just couldn't care less.

I don't like using the word 'desertion'. But I suppose that is technically what happened. For generations, it has been a huge problem for the Legion. The first flood of desertions usually occurs when a new recruit realises that while he thought he was signing up for *Beau Geste*, the reality of Aubagne and Castel is something very different. Some simply cannot take the regime and run for their very lives. For others, it's a gradual process of disillusionment. Barrack life and the dreaded *corvée* can destroy even the deepest commitment to the Legion. And for others it is a multitude of different reasons.

I heard about one Legionnaire who was on duty in Africa and who ran off with a prostitute. After a few months he suddenly decided that he'd made a terrible mistake and somehow made his way back to Aubagne. He tried to hand himself in, willing to do whatever jail time was required to get his Legion life back. But they simply didn't want him. It was apparently too much bureaucratic hassle – there were too many forms to fill in to explain the precise circumstances of his desertion and the reasons for his return. So he was simply told to piss off. It was easier for the Legion authorities to list him as just another Legionnaire who'd tired of wearing the Kepi.

Another guy I heard about was on leave in Paris, met a beautiful young French woman on the very day he was supposed to return to Corsica and simply had to pursue her. Her father had fought alongside the Legion in Algeria back in the late 1950s and treated the young Legionnaire like a returning war hero. But, after a torrid romance lasting several days, the Legionnaire found that he'd overstayed his welcome with the young woman. Her boyfriend was returning to Paris and the Legionnaire's attentions were no longer wanted. So he found himself standing at a Paris train station contemplating whether to return to base and face into a lengthy jail term or go back home. He went home.

For others, it simply seems to be a question of destiny. My friend Mick McCarthy was finishing leave in Paris when he agreed to loan money to another Legionnaire who had drunk his fare back to base. But Mick forgot that he'd already exhausted his own bank account. When he awoke on the morning of his scheduled return to Corsica where the Legion paratroopers are based, he didn't have the money to get to the airport to catch his flight. He rang the base to explain his predicament but the NCOs simply did not want to know. They didn't care that Mick was an incredible Legionnaire and all he needed was help in getting back to Corsica. It was all small-minded bureaucratic bullshit. And in the end, he left the Legion in disgust.

Back in 1996, I simply couldn't commit myself to another five years in the Legion – and then be forced to watch the organisation I loved and respected be diluted to suit some politician's idea about what France's future military needs required. The Legion I loved was being whored to some Paris politician's needs. I found it sickening.

The Legion made me what I am and instilled the values that I treasure, such as honour, loyalty, passion, courage, compassion and humility. People talk about the brutality and beatings at Castel, but I believe some of the most valuable lessons in my life were learned in those Legion bases. The Legion is not easy and it most certainly isn't for everyone. But the Legion made me who I am today and I will never forget that fact. Even now, more than a decade on, some of the people I trust most in life are the men I served beside in the Legion. If any one of them were in trouble, I would drop everything and go to help, no matter where on the planet they were.

I didn't leave without second thoughts. A few of my friends called to see me at my Avignon hotel and, over a few beers, tried to persuade me to come back. I'd already spent more than a few days in the Legion detention blocks so the prospect of some penal time didn't worry me. Hell, you weren't a real Legionnaire unless you had done some time behind bars. No, that wasn't the problem at all. My friends' arguments were quite persuasive: I was almost a Sergeant, so why throw all that away? They also argued that there was the prospect of overseas tours either in Africa or South America, so the trials of barrack life would be avoided for six months or so each time. And they assured me that Guyot would eventually move on, probably by promotion. But I felt that the Legion I loved was changing, and not for the better. Still, the Legionnaires who wore the Kepi and carried on their shoulders a century and a half of pride and courage were the same. And I knew I'd miss these guys.

After a week or so drowning my sorrows in Avignon, I finally decided to head back to Cork. Where else was there to go? The flight back home stood out in stark contrast to my flight out those five years before. The hope, the excitement, the sheer adrenalin – it had all slowly died away. The Legion was changing into something that I didn't recognise and I wanted no part of that change. When I joined I was told that you had to honour what the Kepi stood for and pass the legacy and heritage on to the next generation of Legionnaires. I couldn't honour that commitment by staying. And yet, for the first time in five years I felt as if I'd lost my home. I was absolutely heartbroken and I hadn't a clue what I was going to do in the future. I was twenty-six years old and I was more lost now than I had been at twenty-one. Worst of all, I felt betrayed by the very thing I'd come to love.

The short flight home from Paris to Cork was the longest I could remember since my first assignment to Cambodia, even though it only took about an hour. Images of Castel, Avignon, Sihanoukville, Bosnia and Mont Louis kept flashing through my mind. I was replaying a mental tape of five years in the Legion and not even the drinks I had at the airport could blur the images. I was a proud Legionnaire, but now I'd never again wear the Kepi and that was something that tore at my soul. I'd worked and fought so hard for the right to wear that white hat – was I now throwing it all away? Jesus, what if I had made a mistake! Was it too late to go back, I wondered, as the Aer Lingus plane approached Cork. But I knew I'd made the right decision. I loved the Legion too much to stay and watch it be prostituted.

For quite a while afterwards, I wondered what the Legion's official policy was towards me. When you leave the Legion early it's always something you have at the back of your mind, particularly if you ever intend to set foot on French soil again. A lengthy stretch in a Legion prison is not something to be taken lightly. But a Legion friend told me several years later that I wasn't on any of their official search lists. I had less than four months to serve on my enlistment term and, I reckon, they judged it simply wasn't worth the bureaucratic trouble to bring me back. Perhaps to them I was yet another reminder of the 'old' Legion that they were trying to excise. But I paid a terrible price for my devotion to the 'old' Legion and its traditions. I'd never again shoulder a FAMAS, sing the Legion marching songs or, worst of all, wear the Kepi. And, as the plane touched down on the tarmac of Cork airport in the gloom of an Irish winter evening, I had no idea of how truly painful that price was going to be.

8

A WALKING TIME-BOMB

There's an old phrase which warns that, if you want to make God laugh, just tell him your plans. It's something that, in hindsight, I probably should have taken time to consider in late October 1996. But I arrived back in Ireland with the very best of intentions and plans to forge a new career for myself. I had great ideas. Perhaps I'd set up my own private security company? Surely my training would give me a huge advantage in the security industry? But I had less than €600 in cash saved up from my time in the Legion and, even though I wasn't afraid of hard work, that didn't make life any easier. In fact, after Castel and Avignon, hard work in civilian life seemed to me to be more like relaxation!

But starting my own security business at first seemed like a distant mirage, particularly given my finances. So, I arrived home and decided to drown my sorrows for a while. Leaving the Legion was one thing, forgetting about the Legion was entirely another. I threw myself into drinking with a vengeance and began to look up all my old friends. It was strange, really. I had arrived back in Cobh aged twenty-six but with the mentality of someone nineteen or twenty years old – ready for some serious partying. A lot of the guys I'd grown up with had either moved away or were married and settled down. There weren't many too interested in joining a semi-mad ex-Legionnaire for some serious drinking sessions. Part of me was determined to take up where I had left off almost six years before. But life, as I was about to discover, doesn't stand still.

Without realising it at the time, I quickly began to socialise amongst a younger crowd – lads who weren't worried about getting home in time for their girlfriends or wives. I was trying to make up for lost time. I had gone into the Legion at twenty years of age and now, almost six years later, I wanted to belatedly create some socialising memories of my own. Yet, from this younger crowd came the best friends I have today. It was a mad scene – going out for a few drinks and then disappearing on a session that could last several days. Waking up and, for those first few awful minutes, not knowing where you were or exactly what had happened the night before.

Not surprisingly, after a couple of weeks of major 'benders', my family began to get a bit worried about me. My father thought that a job might help steady me down for a while, so, through his contacts, he arranged that I get some shift work at Cork Plastics, a major employer in the area. I arrived for work and was briefed on what I'd be doing, shown around the factory premises and started my shift. My duties basically involved putting pieces of plastic pipes into plastic bags. About thirty minutes later I looked around me in a combination of boredom and disgust and simply walked out. I didn't know what I wanted from life – but I knew it wasn't this.

So my father then helped get me a job as a security guard on an industrial estate in Little Island near Cork city. At least, I thought, this kind of job might help with my plans for my own security business. I arrived for work and was shown the tiny portakabin that would serve as my hut. Only then did it dawn on me that I was basically there to check the cars arriving at the factory – and absolutely nothing else. It was one step above valet parking. Almost immediately a guy arrived to give me a lecture on my responsibilities and duties, and he started telling me exactly how to patrol the area. I looked around and realised that just two months ago I was entrusted with one of the French Foreign Legion's top combat teams and millions of Francs worth of military equipment. Now I was being told how to properly patrol a car-park.

But, after the debacle at the factory, I decided to bite my tongue and give the job a chance. Walking out on a second job wouldn't sit well with my Legion sensibilities about never quitting. So I lasted for ten days. Then I simply couldn't take any more. I realised that I was probably there only because the plant's insurers demanded it. And simply standing outside the hut, mindlessly watching the cars come and go, was enough to make anyone question their sanity. I couldn't even do roving patrols around the plant because the entrance had to be manned at all times. Throughout, I felt as if I was a non-person. People talked at me, not to me. It was also obvious to everyone in the plant that I was on the minimum wage and they treated me accordingly.

After ten days, I simply quit. I judged it was better to leave quietly than tear someone's head off when they caught me at the wrong moment in a fit of frustration. What I remember most about that job was that I arrived at work angry, I spent the day getting angrier and I went back home after my shift feeling absolutely poisonous. I couldn't understand what was going on. I felt as if the entire world had conspired to make me feel worthless, as if the five years I'd spent in the Legion and in Bosnia and Cambodia meant absolutely nothing.

Pádraig O'Keeffe was proving to be a walking time bomb as a civilian. I seemed to have gone from one extreme to the other – from the discipline and focused life of a Legionnaire to the diversions and dissolution of life on civvy street. My system simply couldn't adjust to the change and I craved excitement even if it meant getting a hammering in a pub brawl. The booze certainly didn't help matters. But I just didn't know any other way of trying to forget my frustrations and disappointments. So I drank to forget and to relax, and the consequences were absolutely disastrous. My father bore the brunt of my anger and, to be honest, it really was a tidal wave of fury.

I got drunk and went down town one night looking to confront him and tell him precisely what I thought of him. I eventually tracked him down to a pub and, when he refused to be baited, I drunkenly staggered out on to the street and began to kick madly at his car which was parked outside.

The Gardaí were notified and a few uniformed officers arrived at the scene but because they knew my father they bowed to his judgement about how best to handle me.

On another occasion I got raging drunk and went berserk inside a pub. One poor unfortunate passed some innocent comment that I took exception to so I grabbed him by the throat and held him, pinned against the wall, hostage-style, as I roared my frustration at the entire world. I've no doubt that anyone in the pub that night thought that it was like a scene from *One Flew Over the Cuckoo's Nest*. Again it was only my father who stood between me and a serious confrontation with the law. I knew I was acting like a lunatic but it felt as if there was a volcano of rage within me that simply had to be let out every so often. Drinking served as the release valve but, in fact, it was more like the trigger in a bomb. I suspect the anger was a combination of fury at what I'd seen done to women and children in Bosnia, frustration at the failure of my Legion career and dejection at having to come home when I really wanted to be somewhere else. After each drink-fuelled blow-out, I'd go through a purgatory of emotions – guilt at what I'd done, shame at what my family thought and anger at how badly I'd let myself down. I was acting as if I was completely off the wall. I knew it too, but I couldn't seem to do anything about it.

By this stage my family had gone from concern to outright terror at what was going to happen next. I was still living at home and, to be honest, I don't think either my mother or father knew what to do with me. It also wrecked my head that here I was, a twenty-six-year-old highly-trained soldier, living with my parents, back in the bedroom I'd used as a schoolboy. Something seemed terribly amiss. As for my parents, well, they simply couldn't understand how anyone could behave the way I was behaving. They kept warning me that I was totally out of control. But they also knew that I obviously had things I just had to talk about before they consumed me like an acid. And, quite rightly, they judged that the sooner I got involved in a project I believed in and diverted my energy and talents into something positive the better for me and for everyone around me.

So I began to open up to my friends about my life in the Legion and, in particular, about what had happened in Bosnia and in Avignon. I tried – sometimes without success – to cut back on what I was drinking and threw myself body-and-soul into launching my own enterprise, Cobh Security Company (CSC). Those few months were a total nightmare for me and they made me question almost every single thing about myself. I felt lost and, at times, it seemed that I was desperately searching for something, anything, worth believing in. I'd been trained to have a certain set of principles and now those principles didn't seem to fit the world that I'd thrust myself back into.

Eventually, by late 1997 I treated my fledgling CSC operation as my life-raft: if I could make that work, then I'd have a purpose and direction. The problem was that, in the Legion, you were constantly told that only the best was good enough and you never, never quit. But in the private security world back in Ireland what mattered most was how much your services cost, not how good you were or what skills you had to offer. What made matters worse was that security contracts were inevitably won or lost on personal contacts and connections and I had very few contacts within the security world in Ireland.

But I looked at the appalling service being offered by some security firms and thought that, even on my worst day, I could offer a service way better than that. They were simply going through the motions. I wanted to offer a modern, challenging security service that might cost a little extra but would deliver vastly more for clients. At first all seemed to be going fairly well. My father had worked for years as a Garda and there was very little he didn't know about the security industry. I ran through my plans with him and said I wanted CSC to offer virtually every type of security service from risk assessments right through to direct patrols and protection. Immediately, he warned me about contract costs and contacts, but I figured that, with the quality of the service I could offer, I'd win more than enough business to get through the critical first few years. After that, I'd expand to handle contracts initially in the south of Ireland before going national. Then, the world was my oyster! But first to secure enough contracts to get through the crucial first year. I'd now invested all of my meagre savings in CSC and, through my knowledge of Cobh, had lined up enough people to do whatever security patrols I was contracted to provide.

I emphasised my experience with the Legion and that I could offer an unprecedented expertise at risk assessment within the Irish security sector. My initial aim was to secure a few contracts from the high-profile industrial and commercial operations in Cork. If these companies already had security cover, well, fine, I'd offer to run an independent risk assessment for them. I'd check what their security team was doing – and, if necessary, simulate everything for them, from vandalism to sabotage and from arson to break-ins. In hindsight, I should have realised that the other security operators must have viewed me simply as a threat. They simply didn't like the idea of a young whipper-snapper waltzing into their lucrative market and showing up the mistakes they were making. So any hope I had of co-operating with the existing players in the security market faded pretty quickly.

It was then that I realised my father was right when it came to contract cost. What mattered for the vast majority of commercial and industrial operations in Cork at the time was not how good my work was or what experience I could bring to bear on assessing their security requirements. No, it was all about how much I cost. Could I under-cut the cheapest quote they'd got from another security operator? A lot of security contracts were awarded purely and simply on the basis of cost. If you were cheapest, you were hired – even if you hadn't a clue how to properly maintain corporate security or your risk assessment measures were totally incompetent. In my opinion, if you worked in the security business you were generally treated as a non-person.

As for recommendations about changing or improving security systems, if it involved any type of additional cash outlay, most contractors simply did not want to know. In most cases, like the guard detail back in the portakabin, the only reason the company was looking for security in the first place was because their insurers were demanding it and they had to comply if their premium payments weren't to skyrocket. It was all pretty disheartening stuff for someone who had been weaned on the belief that excellence mattered.

Admittedly, some firms were different and they were a joy to deal with. I was lucky that one of my first contracts for CSC was from Leahy Brothers in Midleton, a leading East Cork construction firm. It was a great contract to get and Leahys were always a pleasure to deal with. The problem was that companies like that were the exception rather than the rule. I found it all deeply frustrating and, while I was making some money, it was proving very different from what I had imagined. I'd managed to make enough to put a deposit on a house at Rushbrooke Links in Cobh and I was able to drive a decent car.

But, as the months and then years passed, it slowly began to dawn on me that I didn't really like what I was doing. What was worse was the fact that I didn't believe in what I was doing either. I'd been trained to believe that there was only one way of doing things – the right way. Any short-cut or easier route was simply not acceptable. At this point, I even began to get into disputes with clients over which services were necessary and which costs simply couldn't be reduced any further.

I had great friends and, to a casual observer, seemed successful and happy. But this wasn't what I wanted for my life. Deep down, I was still a soldier and, try as I might, I couldn't forget that fact. The anger that marked my departure from the Legion had never quite abated despite the passing of the years and, once again, I found that the fuse for that anger was being lit when I went out drinking. I read in one book on the Foreign Legion about a recruit who asked a Sergeant-Chef what happened to guys when they left the Legion? The answer he offered was: 'They go insane, they go to jail, they die, they become alcoholics or they fight in other people's wars.' They weren't the kind of options that appealed to me as I worked to try to build a future for myself.

So, despite all my hard-won experience, I began to hit booze with a renewed passion. I also began to prove the old adage correct – the more I drank, the less successful my security business was. My heavier drinking sessions once again began to be marked by getting involved in fights. Sometimes, I think I drank just for the adrenalin rush of knowing that I'd get involved in a row. But it was all slowly spiralling out of control once again. One night, after a particularly savage bout of drinking, I got involved in an horrific street brawl. It was probably more of a riot than a brawl and the Gardaí ended up having to wade into the crowd. I don't know whether such rows actually made me feel like I was back in the Legion, but they helped keep me sane in a world where I often felt lost.

Eventually, for my own sanity, I knew I had to change. So I offered CSC – lock, stock and barrel – to my father. I'd had enough. It was a good company and, with hard work and a lot of effort, it could expand successfully. My father had the people skills that the company needed, skills which I knew I so badly lacked. I always said exactly what I thought and sometimes in business that it not such a sensible thing. But I'd laid solid foundations for the firm and there was great potential. CSC was now employing ten people on a full- and part-time basis and, with a little effort, we could expand our order book. Perhaps I had tried to do too much too fast with CSC and, in hindsight, the services I was promoting were probably a decade ahead of their time back in pre-Celtic Tiger Ireland where the over-riding concern was indeed cost. Now, I needed a fresh challenge and, frankly, I couldn't give myself to the firm a hundred percent any longer. I knew my father, with his security contacts, could build the business and he had more than the skills to run the show. So, in the summer of 2002, just five years after I set it up, CSC became my father's concern.

Strangely, I felt free once again. I was never motivated by money, and once I could pay my mortgage, feed and clothe myself and have the odd night out, I was happy. After a few weeks, the first job I took was as a bricklayer's mate. Honestly, I couldn't have cared less what people thought about me swapping running my own business for mixing mortar. The pay was okay – and Ireland was on the cusp of a construction boom, so there was no shortage of work. But it only lasted about three months. The bricklayer was a great guy – he was from Cobh and we'd known each other for quite a while. I needed a temporary job and he was happy to oblige. Yet while we both got on very well, the job ultimately didn't suit me and I decided to move on.

Then I got a job helping out on a waste disposal round

with a guy in East Cork. Basically, I had to help sort the refuse from various skips that we were contracted to handle – putting wood into one pile, metal into another and so on. It was physically exhausting and the pay certainly wasn't anything to shout about. But the lads working with the firm were great characters and, for the month I did the job, I enjoyed myself. I was never worried about doing work that others might dismiss as menial; maybe because of my blue-collar background I always reckoned that there was an honesty about such work. But I also knew I was only marking time. I certainly wasn't going to be doing this work forever and, to be honest, most of my friends suspected as much. I had absolutely no interest in these jobs – all I wanted was to get enough cash to pay my bills and wait until I was ready to do something else.

I was living in Rushbrooke Links in a house that I'd bought from my old CSC clients, Leahy Brothers. I'd bought just before the property boom began in earnest in Ireland so, within a few years, the house had dramatically appreciated in value. It was now worth substantially more than what I had paid for it – despite the fact that once, in a fit of national pride, I decided to paint the entire tiled roof in green, white and gold for an Irish soccer match.

It was a pretty handy asset and, as I drifted from one temporary job to another, a plan began to form in my mind for another business venture. This time, I was going to exploit another one of my Legion interests and I was convinced that it would prove a winner. This time, I'd combine a business with one of my lifelong interests, something that was sure to

keep me focused and involved. And, to prove my commitment, I was prepared to back the venture with everything I
had, including my house. Not for the first time, I was about
to take a chance.

9

BACK TO BUSINESS

After all that had happened, it probably wasn't surprising that I decided to rest my remaining hopes on fulfilling an old dream – establishing a dog-training centre catering for both pets and guard dogs. For as long as I can remember, I have loved dogs. To me they truly are man's best friend. In the Legion I'd wanted for a long time to go on the K9 dog- training course, only to be frustrated at every turn. As it turned out, my dreams of undertaking the K9 played an indirect part in me quitting the Legion. I'd forgotten about the whole idea for almost six years until, after I'd had my fill of my private security business, I was casting about for something else to do. But then I discovered through a contact that a US company was offering a highly regarded dog-training course.

The firm was based in Columbus, Ohio, and its qualifications were accepted throughout the world. In fact, the course was so good that a lot of international police forces used the course to establish their own dedicated K9 units. I loved working with dogs. Maybe the Ohio course was precisely the change in direction my life needed?

I decided to take the chance. I hadn't much ready cash to fund what was likely to be a very expensive course in Ohio. As well as the course fee, I'd have to support myself in Columbus for eight weeks and I also fancied a two-week break after the course concluded. So I decided to sell my house and use the proceeds to fund my plan. I made a tidy profit from the house sale even after clearing what was left of my mortgage. I had by now invested all my savings – and my faith – in the venture. Now, a bit like my journey to Aubagne from Cavan all those years before, I had nothing left to lose.

The course in Columbus was one of the world's best. If you completed the eight weeks here, you could operate a K9 facility anywhere. But it was far from easy. There was no guarantee of a completion certificate, and, if you could not progress with the required tasks with the particular dog you were assigned, you were dropped from the programme. What also took me by surprise was that the course was Legion-like in its intensity. We were operating up to sixteen hours a day – and you were expected to practise the drills and skills with your dog in your own time.

Crucially, the most important lesson was the ability to relax and not rush the process. Patience was the golden key to working with dogs and had to be shown at all stages of the process. If you rushed it you jeopardised all the work you had put in and everything you were trying to achieve. Even now I smile when I realise that that lesson could as easily have been applied to me as to the young German Shepherd I was working with. I needed to refocus my life and learn to apply my skills patiently and consistently. With CSC I'd wanted everything done yesterday and that kind of impatience was a recipe for disaster. The old saying goes that 'Patience is mother of virtue' and I was about to find that out for myself.

I worked slavishly to practise what I was learning with my assigned dog and together we passed all the various tests set by our Columbus supervisors. It was hugely rewarding and I found real satisfaction in my work for the first time in years. I simply loved training the dogs. It was incredible what the animals could learn and how a good trainer and dog could operate as a seamless team. I saw a huge potential for such skills back home, particularly given the large number of pets traditional in Ireland. I wanted the centre I would set up to cater for everyone – right from the mongrel that was the family pet through to the German Shepherd or Doberman that would be undertaking security patrols.

I'd dreamed up the name 'DogTags K9 Training Centre' for the complex and I wanted it to cater not only for dog-obedience training but also for grooming and kennelling. It would be the first fully-integrated complex of its kind in Ireland. I reckoned that my centre would particularly appeal to the security industry which still had a heavy reliance on old-fashioned dog-patrol work. Despite the advent of an extraordinary range of technology within the security industry, there was nothing quite as effective as a well-trained security dog, both to detect threats and to act as a deterrent. I was actually gutted when the eight-week course finished, though it was a great consolation to have two weeks to myself to enjoy the US. But I was really confident on the flight back to Ireland that I'd finally found a niche where I could combine something I loved doing with earning a decent income.

Arriving back home, I set to work at a feverish rate. I'd located an ideal site for the centre at Marlogue, a beautiful little inlet just down the harbour from Cobh. The landowner was very keen on the idea and generously offered me a lease deal whereby I could make payments once I had the venture up and running. But I still had to find the money for a large shed to be erected on the site, the installation of proper fencing, construction of a special training course, not to mention all the other hidden costs that seem to plague every new venture. Every single day seemed to bring a new cost and the paltry sum that I had saved up to fund the venture now began to disappear at an alarming rate.

I tried to ease the financial pressure on myself by applying for a government grant. But there was none available. I was creating the first dedicated dog-training centre in the entire State yet there was no scheme available to support me. I was angry that millions were being offered in tax relief to the equestrian industry and that stud farms didn't even pay tax on their earnings from stallions! I contacted a local politician and, after making inquiries, he confirmed the bad news. There were numerous grant schemes open to me if I was working with horses, but not a single one for dogs. So much for a level playing pitch.

Sadly, a bit like my dreams for CSC, DogTags K9 seemed fated not to succeed. The idea never really caught on in

Ireland and, while the dog-training school in Cobh set new standards and impressed our customers, we simply failed to secure enough business to remain viable. I tried everything I could think of, from advertising to word-of-mouth endorsements. I was absolutely convinced the business was there – the problem was it was not coming to our school in the volumes needed to keep us solvent.

I worked hard but, to be honest, I still didn't get the excitement or satisfaction from running the centre that I had got from my days in the Legion. I liked the centre, but I loved being a soldier much more. In the end, my dog centre was operational for about a year, but by this stage it had consumed all my savings and it was only a matter of time before it shut down for good. I have to say I was heartbroken – it was a venture I really believed in and, even now, I reckon it has a niche in Ireland. I was very proud of the centre. We really had made a great job of it.

But I was now completely out of ideas and fast running out of money. The two areas where my gut-instinct had told me there was a security niche had both effectively proved failures. And setting up the dog-training school had also left me all but homeless. My savings were virtually exhausted and my options were running out. I pondered over the irony of being the most highly trained security operator on the dole queue. For the first time in years, I really did not know what was next.

Then, in the midst of all this turmoil, my past caught up with me and offered me a way to put things back on track. Another former Legionnaire, Mick McCarthy, had completed a specialist bodyguard course in Norway and had made some good contacts over there. Out of the blue in early 2004, he got a phone call from a few Norwegian guys who were interested in contracting for some security work in Iraq. Would Mick be interested in doing some work for them? They travelled over to Ireland to sort things out with Mick and he recommended me for some work as well. I sat a short interview in a Thai restaurant in Cork city and then ended up hitting the town with the Norwegian guys in tow. It wasn't the first drinking interview I'd ever had, but it changed my life like none of the others.

The Norwegians were really impressed with our Legion credentials and wanted us on board in their sub-contract work. Initially, Mick and myself thought that it would be a few months before we'd hear anything concrete from Oslo. But, to our surprise, they were back in contact within a few weeks to say that things had progressed really well and they wanted us in Heathrow to meet up with the team bound for Iraq in a few days' time. It was almost surreal. One minute Mick and myself were watching the unfolding tragedy in Iraq on the news – and then suddenly we were going to be part of the security apparatus trying to protect Western civilians from the savagery of the insurgents.

I'd few reasons to stay. The dog school was folding and it didn't have enough business on its books for me to sell it as a going concern. I had few prospects of landing a job in Ireland that I liked and it seemed that civvy street wasn't a happy destination for Pádraig O'Keeffe. After all that had happened over the past eight years, I'd finally got the message. I might picture myself as a businessman but, deep down, I was still a soldier. It's what I did best and, so far in my thirty-four years, it was the thing that gave me a feeling of satisfaction and fulfilment. I'd no compelling reason to stay at home either.

In fact, the only hard part was telling my family where I was heading. The Iraqi security contract was confirmed in early 2004 and just weeks later, while we were still waiting to take up our jobs, the world news was dominated by the horrific execution of Nick Berg, a twenty-six-year-old US businessman who was kidnapped in central Baghdad and then beheaded in a gruesome manner on 11 May 2004. The tragic plight of that young man underlined precisely the savage nature of what was now happening in Iraq. Suddenly the image of a lonely figure in an orange jumpsuit seemed to illustrate man's gruesome inhumanity to man. Iraq was a conflict where what few rules had existed were now being thrown out the window. Not surprisingly, my parents were absolutely appalled at the idea of me getting involved in the security industry there. My mother had been sickened at what happened to Berg and, I'm sure, was now fearing the worst for her only son. There was no way she wanted me in Iraq and she tried, in vain, to talk me out of fulfilling my contract.

It hurt to see my parents so worried, but I felt that this contract offered me the focus that my life had lacked for too long. As far as I was concerned, the past eight years had been soul-destroying. Iraq was the only realistic option I could see. I also think that, deep down, my parents understood that this was what I excelled at – I was a soldier. I was trained for precisely this type of situation. As far as I was concerned, there were people who needed protection and I had the skills to help keep them safe.

But those skills were very rusty after almost eight years in Ireland. To be honest, what made me most nervous heading out to Iraq was the fact that I knew I needed to work hard to get my skill levels back to what they were with the Legion in 1996.

Luckily, I'd never let myself get too badly out of shape. I was nowhere near the frightening fitness levels I'd boasted after Castel, but I was fairly fit compared to your average thirty-four-year-old. A regime of training and drills would burn off whatever excess fat I'd put on. I knew I was going to pay dearly for those pints I'd had over the past few months!

My biggest concern was to do with my shooting skills. I hadn't held a rifle in almost eight years, let alone gone through a regime on a target range. I knew I needed to train hard to get my accuracy and response times ready for a combat situation – because I knew that even the slightest hesitation or second's delay could have fatal consequences. Again, the Legion training helped. I was always pretty comfortable with the infantryman's drill and I knew precisely what I had to do once I was 'in country'. I'd be fine as long as the guys I'd be working alongside would allow me the time to work off the rustiness until I was ready for the streets of Basra or Baghdad.

So Mick and myself packed our bags for Iraq. We flew

from Cork to Heathrow and met up with the rest of the Norwegian team there. They were a nice bunch of lads and great company, if a little gung-ho. I'd viewed their video presentation to the English security firm, the Hart Group, who were responsible for all the contracts in Iraq, and it was pretty impressive. There was lots of military hardware on display and the guys seemed pretty competent.

We were given a briefing session by the British contractors on what the work in Iraq involved and precisely the type of security situation on the ground. The contractors were very efficient and at no stage tried to conceal anything about Iraq and its dangers. The bottom line was that it was very bad, and was getting worse by the day. The deaths of security contractors generally didn't make the news headlines in the same manner as the deaths of US or British soldiers. But the private security contractors in Iraq were clearly paying their own toll in blood for getting the supply convoys through.

Even though I was glad to be back in an industry that I thought I'd left for good, I have to say I had misgivings about the Iraqi contract. There is a world of difference between going into a combat situation as part of an army like the Marine Corps, Foreign Legion or Royal Marines and being a private security contractor. A soldier knows he is a small part of a very big machine and there is comfort in knowing that that entire machine is behind him. It's very different for private security contractors. I knew that in Iraq I'd largely be responsible for my own safety. If the worst-case scenario came about, I knew there was little likelihood of the cavalry coming to my rescue.

Though the contracting companies looked after their personnel very well and there was a tremendous bond of comradeship between contractors, we all learned pretty quickly that when a US or British soldier was shot he was a casualty, but when a private security contractor died he hardly registered as a statistic. The most appalling fact of life for private security contractors was that, in most cases, if they were killed in an ambush by insurgents, their bodies were hardly ever recovered. In one case, the family of a slain security contractor even received e-mails gloating over the fact that his body was missing and unlikely to be recovered. 'It's all about fucking with your mind. This war is being fought as much with TV cameras and the Internet as with assault rifles,' one briefer told us.

But I still walked out on to that flight to Kuwait and felt that my life, after eight years on pause, was finally being lived again. I'd always wanted to be a soldier and here I was, after eight years on civvy street, back where I reckoned I belonged. The flight itself was uneventful and, after a few hours of routine processing through the airport at Kuwait City, we joined a convoy bound for Basra in southern Iraq which was part of the British-controlled sector. It was a Shia-dominated region and, to date, had witnessed nothing like the insurgent attacks that had marked the US sector farther north around Baghdad, Kirkuk and Mosul. But we were no different from the other civilians that made up the convoy – we weren't given weapons and our security depended on the tanned contractors who ran the protection cars that whizzed around us.

As the convoy set out from Kuwait towards the Iraqi border, I felt alive. I was doing what I had been trained for and suddenly everything seemed to make sense again. Given all that had been happening in Iraq over the previous months, I probably should have had a feeling of dread as the safety of Kuwait City faded in our rear-view mirrors. I knew I was heading into a world that any sane person would have fled from. But, to be honest, I was pumped up with adrenalin and felt the same way I had back in those early days with the Legion. Yet I was aware that to survive Iraq I needed all my skills honed the way they had been a decade before and as we approached the heavy fortifications of the Iraqi border, I knew there was plenty of work to do.

10

HOSTILE TERRITORY – BASRA

Before I'd even arrived in Basra the signs were ominous. I already knew this was going to be very different from anything I had experienced in either Cambodia or Bosnia. But the extent of the carnage on the ground took us all by surprise, from the burned-out wrecks of ambushed vehicles through to the bullet-riddled road signs.

Well, most of us were shocked – though we had barely arrived in the security compound in Basra when a few of the Norwegian lads had started to raise eyebrows. They were really decent guys, and had got on very well with all of us on the trip out. But, within a couple of hours of arriving, I spotted a few of them walking around dressed like extras from a Rambo film. They had changed from their travel clothing and were now wearing bandanas, special combat fatigues, black T-shirts, wrap-around sunglasses and one guy even had a set of Ninja throwing-stars hooked to his belt. I stood in disbelief as they paraded around the security compound and started taking snapshots of each other in various dramatic poses. It was over-eagerness, but also indicated a complete lack of experience of how to behave in hostile territory.

If I wasn't so shocked I think I would have found the scene hilarious. Here we were, in probably the most dangerous place on earth where one mistake could get you killed, and these guys were behaving like they had arrived in the big boys' version of Disneyland. I began to suspect that these lads lacked any serious military experience – they were willing, but just didn't have 'in-field' hours. I've no doubt they could have done the job, but the risks were very high.

Luckily, the security director was a Special Air Service (SAS) veteran and could sniff trouble from one kilometre. You only had to take one look at Sam G. and you knew that he could walk the walk. Anyone I ever came across who had been associated with the SAS knew their stuff and Sam was no different. He was cut from precisely the same cloth as Santos and Gercer – and I instantly knew my new Norwegian friends were in trouble. Typically, they completely failed to understand the concern they were causing and continued with their photo-antics.

Less than ten days later, the photo-snapping Rambo-wannabees were, without exception, all gone, and only the few capable ones, like my friend Gere, were assigned to security teams. Apparently, those guys had had nothing like the military experience they had boasted of. They were more like adventurers and weekend warriors than experienced security contractors and that was enough to earn them an instant return flight to Heathrow. In applying for the job they had made a fantastic video presentation with borrowed vehicles and equipment! The problem was, it wasn't worth a damn in the killing fields of Iraq.

One of the senior British operators admitted afterwards that it had been a no-brainer decision to send them back. 'Those fuckers would have gotten themselves killed – and that's their look-out. But they'd probably have taken a few of us with them,' he grimaced. Sam and his team knew from the start who had competent military skills. I went to him straight away and asked could I spend some time on the firing range. I told him the truth that it had been a few years since I'd been in the Legion and held the working end of a rifle. He knew straight away what I needed and he put Mick and myself through a pretty intensive training regime with the Kalashnikov on the range. Ammunition was never rationed for us and we got all the live-fire practice we needed to work off the rustiness.

They were a pretty tough bunch in Basra. The security personnel were almost all English-speaking, with former soldiers from Wales, Ireland, England, Scotland, Australia, South Africa and America. But there also a few French and Serbs thrown into the mix. I learned immediately that all these guys had served with pretty good military outfits and were no strangers to combat. One British guy had served with the Parachute Regiment and sounded like he had just stepped off the parade ground at Sandhurst (I later found out that one of his distant ancestors was the Duke of Wellington). He was the absolute model of a British Army soldier – experienced, cool, totally professional. You knew after just five minutes with him that he could not only 'talk the talk' but had also 'walked the walk'. This guy knew his business and was a very serious operator. Once you got to know him, he was a great guy and a good colleague. The whole operation was directed by Sam and another ex-British Army guy called Ian.

I found the entire Hart operation like a mirror-image of the Legion. There were so many nationalities mixing together – but you also knew that the guys around you knew their stuff. Some of these characters were the best soldiers I'd ever served alongside. They had experience of everything from special operations to reconnaissance missions and from bodyguard duties to urban warfare. And, for our protection details, we would operate exactly like a highly-skilled small infantry unit. We'd have assault rifles, side-arms and one heavy machine-gun per team. But, unlike an infantry unit, our aim was to avoid trouble if at all possible. Our mission was to protect the people and materiel in our care. But, if trouble found us, we'd be ready for it.

Sam told Mick and myself that, for the first few weeks, we would float between the various security teams to ascertain our skills and which duties suited us best. But, for the most part, we'd be working on convoy protection and personnel security. For convoy work we needed to learn the risks of Improvised Explosive Devices (IEDs), and how to spot and avoid them; the problem was that spotting them was a near impossible skill. But we'd also learn how to maintain a low profile while doing our job. As well as that we'd learn the precise 'hot-spots' around Basra and southern Iraq that were to be avoided at all costs.

Tragically, the precise extent of what faced us was underlined pretty quickly when one Legionnaire I'd served with fell victim to the insurgents not long after I arrived 'in country'. Nick Pears had been a great mate since we'd gone through Castel together twelve years before. I knew that he had gone to the 2nd REP on Corsica while I had gone to the 6th REG – and, unfortunately, we had lost contact. The Brits have a love-affair with the Legion parachute regiments and Nick was no exception. But, like myself, he'd become increasingly frustrated by the way the Legion was being run and the way the old Legion we loved was being transformed. So, he'd had enough and left.

From what I later heard, Nick had drifted into security work back in Britain and had agreed a contract to work in Iraq sometime in early 2004. I was in Iraq a few weeks when I got word that he was also 'in country' working as a contractor. But he was up north in Baghdad. Before I could even start to make plans to link up with him, he was dead.

He had been killed by insurgents during a major ambush staged just as the convoy he was protecting was crossing a bridge on 3 January 2005. The insurgents were learning their trade with deadly efficiency, particularly when it came to the use of IEDs. These were very difficult to protect against, particularly the roadside variety, and only the most heavily armoured vehicles, like main battle tanks, were immune. If an IED was strategically placed and had a sufficient explosive charge, it was lethal to lightly-skinned vehicles like cars, jeeps and even light military vehicles like Land Rovers or Humvees. The insurgents had even evolved different types of IEDs, some of which had copper cores with fuel, which were particularly lethal for non-armoured security vehicles. The explosion would pierce the outer skin of the vehicle and then end with a high-temperature fireball inside the car or van.

In Nick's case, the IED did its work with terrible effect. The explosion had wrecked several vehicles in Nick's convoy as it passed but, unfortunately, it had also blown the car carrying Nick against the bridge's guard-rail. Under the force of the impact, the saloon careered right through the barrier and plunged down to the ground below. Nick apparently survived the initial explosion but died in the impact of his car coming off the bridge and striking the ground. It was a harsh lesson that sometimes all the skills in the world cannot protect you if you're in the wrong place at the wrong time.

It was perhaps lucky for me that I was based in Basra throughout my first tour as it wasn't anything near as bad as what was happening farther north in Baghdad. Still, I didn't believe in taking any chances and, as I established myself with the security team, I stuck rigidly to the rules of my Legion training and the form of defensive deployments I had learned there. I quickly got back to my training discipline and everything was triple-checked right down to the smallest detail. Whenever our vehicles had to stop en route, I set up an immediate defensive perimeter complete with firing positions. I was playing it by the book, even if some of my contractor colleagues thought I was a bit overzealous.

The overwhelming majority of guys I worked with were decent blokes – experienced, good company, loyal comrades and great to work alongside. There were a few exceptions, though, in all my time in Iraq there was only one guy I really disliked, a Scot who decided that my Legion training was a bit excessive and tried to embarrass me and make me cut corners, as I saw it. Once, when our convoy had to stop because of a problem with a truck, I began setting up a defensive perimeter. This guy, who apparently had served in one of the Scottish regiments within the British Army, walked over to me and told me to take it easy. I looked at him, shrugged and silently kept on with my precautions.

'Hey, Paddy,' he joked, 'not everyone here is dangerous. They don't all want to kill you.' When I ignored him and focused on my perimeter work, he laughed out loud, walked over to the roadside and started waving at every passing Iraqi vehicle. Naturally, a few Iraqi drivers good naturedly waved back. The Scot turned, clearly satisfied with his public display and, with a smirk on his face, said: 'See, Paddy, they're not all dangerous. So you can lighten up a bit.' I was new to Iraq so I kept my mouth shut. I contented myself with the thought that I hoped he would still be able to wave when a driver pointed an AK-47 directly at his head.

After about a month, I was finally happy with my skill levels and I was now as familiar with the AK-47 as I had ever been with the FAMAS. In theory, we could have insisted on our own weapons. And while there were M16s and even German-made Heckler Koch assault rifles in circulation, our security team preferred the venerable Kalashnikov

AK-47. The AK had traditional Russian values: it was rugged, it worked in all conditions and, while it wasn't pretty, it got the job done. Even though it was designed in the late 1940s, the AK still had a formidable reputation. It was lethally reliable, and, in Iraq, the insurgents used nothing else. In fact, almost everyone in Iraq seemed to have an AK-47, sometimes even kids aged about fourteen.

So, if we used the same kind of weapon we could, at least in theory, re-arm or re-load from local Iraqi supplies. We also went with the original AK which fired a 7.62mm round. The Russians had updated the gun to fire a higher velocity 5.45mm rifle round, and simply renamed it the Kalashnikov AK-74M. But I personally felt the 7.62mm round was more effective. It had greater stopping power and, in the nightmare security situation in Iraq, that was paramount. Modern armies now tend to favour the smaller calibre 5.45mm bullet because it travels at greater velocity or speed. It also offers other advantages in terms of medium-range accuracy. More often than not, a high-velocity 5.45mm round will exit a person's body after striking – and the 'suction' it leaves behind in the wound can then prove as lethal as the damage caused by the bullet itself. But the 7.62mm round, with its higher calibre or size, travels at a lower velocity or speed, but, because it is bigger and slightly heavier, hits its target with a greater 'punch'. To use a rather unscientific phrase, *immediate stopping power* tends to favour the old 7.62mm round.

I also had my own modifications, thanks to a great friend who works with the German arms company, Heckler Koch.

Firstly, I'd got a special extending sling for the rifle. This meant that I could sweep the rifle from a 'rest' to a 'firing' position simply by increasing pressure on the elasticised strap. It saved only a few micro-seconds in the response process, but those few micro-seconds could make the difference between life and death. My buddy had also made up some heavy reinforced-steel bracelets for me. These weren't a fashion accessory – they were designed to allow me to smash my way through a glass window or even plexiglass, if I was trapped inside a vehicle, without shredding my arms.

I also sourced a Russian-made Makarov 9mm automatic which I wore in a holster at all times. While a pistol isn't much use in an ambush, it's easier to use in a confined space like a car – and it also offers the ultimate insurance policy against ever being forced to wear an orange jumpsuit, everyone's nightmare scenario. I had to buy the Makarov myself, but I didn't mind because of the peace of mind it offered.

The Iraqi security contractors who worked alongside us were lovely blokes, all family men who were simply trying to earn a wage and put bread on the table for their wives and children. And they could source almost anything for you, particularly when it came to weapons and ammunition.

I learned fast in Basra, which was just as well because the savagery of Baghdad and the Sunni Triangle was gradually spreading down south. One of my first convoy escort missions was to the north of Nasariyah. As I sat in a nondescript security car escorting the heavy artics, I was struck by the number of burned out wrecks by the roadside. From trucks to jeeps and cars, there were dozens of these old, charred hulks lining the roadside. Each one represented an ambush, and, I thought, most were probably the grave-markers of security contractors.

Sam and his team ran a very cautious outfit. We tried to vary our routes as much as possible and we never advertised in advance what we were carrying or where we were going. We used battered old cars as our security vehicles to make them look as much as possible like ordinary Iraqi civilian cars. And we learned to anticipate the tell-tale signs of an ambush – from a BMW with darkened windows shadowing the convoy at the rear to a dead dog or donkey lying right by the roadside. Several convoys had come to grief by insurgents simply taking a dead animal, packing it with explosives and leaving it by the roadside until a Coalition or civilian convoy passed by. Nothing was sacred. I listened and learned, and discovered that nothing could be taken at face value here in Iraq. Even a pile of rubbish by the road had to be treated with maximum caution because you never knew if there were explosives concealed within it.

I quickly found that most of my time was spent on personnel protection. We provided security for everyone from engineers to oil experts. They simply couldn't function in Iraq without protection – and that was my job. We learned how to use the four-car security detail to maximum effect. This entailed one car carrying the client, a scout vehicle to secure the approach route and two 'delta' cars which were entrusted with dealing with any trouble if it arose.

The allocation of security missions came from the tasking

office within our Basra compound and they liked to have set teams work together – personnel who knew and respected each other were more efficient and much better suited to the demands posed in Iraq. Needless to say, some of us got on better together than others. Unfortunately, despite the fact that Mick was one of my best friends, we were never assigned to the same security team. Initially, I worked with a team led by an English guy. He was nice enough and introduced himself by saying that he wanted any problem brought directly to him so he could sort it out face-to-face. He said he wanted his security team to be a tight unit. That's a good start, I thought. But it was to prove very different in practice.

After a few weeks, I discovered that this guy liked to leave the hard work to others. While the rest of the security team would be checking the cars, inspecting tyres, oil and coolant levels as well as reviewing the proposed route for the day, he would be in his tent playing computer games. When everything was ready, he expected to be called on the radio and then we'd go about the day's business.

However, about two months into my Basra tour we ran into logistical problems with a proposed convoy. One of our security cars had been giving problems, which was hardly surprising given its age and battered condition. We decided to replace the car, and then, typically, the replacement vehicle also started giving problems. The entire security team was working flat-out to try and get everything ready and double-checked for our scheduled departure slot and pick-up with our client. Then our so-called leader rang on the radio and demanded to know what the problem was.

I told him and, when he gave me an earful about not wanting to be late for the client pick-up, I blew a fuse. 'If you've a fucking problem with us being late then put down the PlayStation, get off your arse and come out here and help us,' I snarled. I now learned that his notions about wanting an open and honest relationship with his team members wasn't all it was supposed to be. He immediately went to his superiors and told them he wasn't happy with my attitude and felt he was compromised by having a team member who didn't fully accept his authority. He argued that this ultimately impacted on the safety of the client. 'I just can't trust Paddy on the road any more with his attitude,' he whined.

I just couldn't believe it. After ten years away from the bureaucratic bullshit that was destroying the Legion, here I was again at the receiving end of some pocket Napoleon. The senior security team directors pulled me aside and said they were transferring me to another security detail for the day. 'Look, Paddy, it's nothing personal,' one of them explained. 'Don't make a big deal out of this. It's better for everyone if we play it like this.' I was fuming and, in different circumstances, would probably have barrelled into the guy. But this was Iraq and there was no room for showboating or letting personal grievances undermine overall team security. I accepted what I had been told and went to another security team. But about three weeks later the guy was demoted after the other team members were discreetly queried about his behaviour.

Ironically, the transfer was quickly followed by one of my first tastes of ambush action in Iraq. My new security team leader was a giant American. Together, we were to provide a security escort for a heavy lorry bringing an industrial power generator from the Kuwaiti border into Basra. The huge generator weighed several tonnes and cost around $250,000 (around €200,000). The problem was, they were painfully slow to move and made a mouth-watering target for the insurgents.

We collected the generators at the border and when we were about 40km from Basra, the trouble started. First, an old Ford saloon swept up behind the convoy and the driver started to play a game of cat-and-mouse between our security cars and the artic. He would keep trying to get between us and the trucks and we would keep forcing him back. Then, at precisely the wrong moment, the heavy artic broke down. We had barely enough ground to a halt when I realised that the Ford had screamed to a stop behind us and three guys were getting out, with the driver staying in place.

Without thinking, I levelled my AK and, when I spotted one of the guys with a rifle in his hands, I emptied an entire magazine at the cab. Paul, who was my detail partner, followed suit. Our quick response to neutralise the threat had clearly taken our attackers by surprise – and I reckon we took out all four. But the truck couldn't have broken down at a worse spot – we were on an open road separated from a small village by a stretch of open waste ground. Our civilian driver was whisked into one of our 4x4s by some of our Iraqi security guards. But the villagers nearby obviously decided that the generator was a gift from Allah and they quickly joined in the attack. I could see the tell-tale dust 'puffs' of rifle rounds falling short of our position.

Then, in the distance, I heard the sound I'd been praying for – an approaching convoy. Our combined firepower would be more than enough to fight off the insurgents. But, as soon as my hopes were raised, they were dashed. The convoy was made up of Coalition military vehicles with a full US Army escort. Incredibly, they accelerated when they saw the fire-fight we were engaged in and drove off. I still recall the shock I felt when a US soldier, manning a .50 calibre heavy machine-gun, turned his back on our position as his convoy sped past. Just a few rounds from that .50 calibre would have ended the ambush, but the US convoy simply drove away.

By now our position was becoming increasingly dangerous. If we were outflanked by the insurgents we risked being slaughtered. Our security leader deemed the potential threat to be too great to stay. So we ordered everyone out of the area, leaving the crippled truck and its expensive cargo behind. Our security cars opened up with heavy fire to cover the retreat and, even still, we were very restrained in what we shot at. The insurgents or hijackers would fire at anything that moved – man, woman or child. But we only shot at armed personnel facing us and, if those insurgents were surrounded by civilians, we did our level best not to injure non-combatants. As far as I was concerned, it was the soldier's code.

We had only travelled four miles when we met a British Army patrol and, while they were sympathetic to our plight, had their own tasking orders and could not support us if we re-engaged to protect the truck and its cargo. But, after conferring, we decided to return to the truck to see if we could salvage the situation.

We screamed back down the road, brought our jeep to a halt amid burning rubber, and re-engaged the hostiles. Myself and three Iraqi security contractors advanced in patrol-formation, firing in the air to warn the villagers away from our truck. By now the insurgents had become mixed in with the villagers around the stranded truck. The rest of our security vehicles roared up and the insurgents began to move back, obviously worried about what forces were now facing them.

I scanned the area where the firing was coming from and caught a figure, armed with an AK, running back towards the village. I immediately took aim and, just as I was about to send a volley at the running figure, I realised that it was only a teenager. He couldn't have been more than fourteen or fifteen years old. Part of me wanted to kill him, but, again, another part didn't want the life of a teenager on my conscience. The terrifying thing about conflict is that these decisions are made in seconds – and I decided to aim wide. I peppered a few rounds close to the teen but deliberately wide of him, and he immediately dropped the AK and ran for his life. It was a great recovery by him, but he'll never know how close he came to being killed that day.

I scanned the place where I had fired at the teen. Sure enough, I found the AK-47 that he had dropped and, on examination, I found that the rifle had been fired and even had a round in the breach. That Iraqi kid may only have been fourteen or fifteen but he was already trying to kill. Yet it didn't change the fact that I didn't want it on my conscience that I'd shot a teen through the back. But I also knew that, had the kid turned around and tried to return fire when I had deliberately fired wide, I would have killed him on the spot. I'd be a liar not to admit that the thought crossed my mind afterwards about whether that teen would pick up a replacement AK and try to kill another security contractor, Iraqi policeman or US soldier.

I mentioned the incident to one of the British contractors and he was familiar with the problem, but he laid great emphasis on the fact that the kid *had* fired a rifle and was trying to kill. 'Paddy, what if he goes on to kill one of us in a few weeks' time? It's not an easy call to make, but if you pick up a rifle in that type of firefight it's either kill or be killed,' he advised. I knew he was right, but it didn't ease the turmoil in my head. As in Bosnia, I didn't want the life of any child on my conscience, and I didn't want to shoot anyone in the back, it just wasn't part of the soldier's code that I'd believed in for so long.

But it was my first taste of real insurgency in Iraq and it slowly began to dawn on us all that things were going to get a lot worse. If I thought Bosnia was bad, well, Iraq was about to re-write the horror story. The IED attacks were increasing in both their frequency and sophistication. The so-called 'nuisance' attacks with sniper fire, rocket-propelled grenades and mortars became an almost nightly occurrence. I also began to realise that the American units in Iraq were stretched to the limit.

My first trip home was a welcome break. My family were relieved to see that I was okay but I didn't go into too many details about what it was really like on the ground in Iraq. I knew they were watching the news each evening on the TV and I didn't want to add to their worries. It was also great to catch up with my friends in Cobh and I let off steam with them for a few days; after the tumult of my first tour in Iraq, it was precisely the wind-down I needed. It was a very strange feeling – being fully aware of the dangers in Iraq and yet relieved to have rediscovered a military life that I had thought was lost to me. I also took pride in the fact that, despite having left the Legion so many years before, I'd quickly shown in Iraq that my training had not deserted me. I was now a pretty skilful operator. I could scan the road-ways for danger signs, plot a convoy route and direct an ambush response. My marksmanship had also improved dramatically and while I wasn't back to my Castel/Avignon peak, I wasn't far off it. I knew I was respected by the senior security contractors for my skills and judgement, and I guessed that, having proved myself, a return contract would be offered by Hart. I wasn't wrong and, just a couple of weeks after I'd come home, I was offered another tour of duty in Iraq.

The intensity of the conflict and the incredible pressures it exerted on security personnel were fully recognised by employers. The British firm I worked for operated on a rotation basis of twelve weeks' field service after which you went back home on leave. When you flew home from Iraq you were listed on the bottom of the 'leave pool'. The longer you remained on leave, the higher you ascended towards the top of the pool until, on reaching the top, you returned to Iraq for your next tour of duty. The length of leave could vary according to personnel numbers, the security situation in Iraq and the work being secured by the contractor.

The contractors knew the pressures and they also knew that, for many, a security contract in Iraq could end in death or serious injury. They did everything they could to make our days off in Iraq as tolerable as possible, which, admittedly, was an almost impossible task. Critically, the food on offer during my first tour in Basra was fantastic, as good as in many hotels I've eaten. The cooks, mostly Jordanians, varied the menu and the quality was always high. The company also tried to provide radios, CD players and televisions to help us wind down after a stressful period of duty. Some of the younger contractors had brought along computer game consoles and lost themselves in these during their time off.

Personally, I liked to keep an eye on how Manchester United were doing and watch any live football that I could. But I also smoked far more than I know I should – it was a crucial tension-release valve for me in Iraq. One of my favourite ways of bleeding off the tension was simply to sit on the steps of the canteen or the entrance to my tent, smoke a few fags and chat over the events of the day. In Basra, I often linked up with Mick, if he was around, and we'd smoke and chat.

I frequently met up with old Legion comrades in Iraq.

One evening I was sitting having a chat with Mick in the darkened doorway of the tent when a hulking figure strode by. Because of the security situation, lighting was kept to a minimum so as not to given the insurgents something to target. The giant figure stopped and asked in an Australian accent: 'Give us a light, mate.' I instantly recognised the voice as that of an old acquaintance of mine from the Legion. Without hesitation, I replied: 'Fuck off and get your own light, Skippy.' I saw Mick looking at me strangely, thinking that a brawl was inevitable. But the giant stopped, peered into the darkness and said: 'Jesus, Paddy, is that really you!' A few minutes later we were having a smoke, sharing a drink and catching up on old times.

Every so often we'd get a case of beer or a bottle of whiskey and have a few drinks, but you couldn't over-indulge because you never knew when you'd be needed for a patrol or convoy-escort mission. Iraq was definitely not a place for driving around holding a loaded Kalashnikov and nursing a hangover.

The three things that no-one could do anything about was the inferno-like heat, the sand and the plague of flies. There was simply no escaping their relentless attention. Some people who've been in Spain on holidays during a heatwave think they know what high temperatures are. Believe me, they don't. Working in Iraq during the summer months was like making a blast-furnace your home. It was unbearably hot everywhere, and going anywhere inside a car was torture. It got so hot that you couldn't even directly touch the exposed metal of the car or you'd burn your skin. The British

Army issued new desert boots to its troops and the soles began to melt in the ferocious heat!

The irony was that, just as you were learning to cope with the heat, the summer ended and the Iraqi winter set in. The winds sweeping over Mesopotamia made Iraq resemble Siberia and, while you might laugh at the instruction to bring warm clothing in July or August, by December and January you were grateful to be able to wrap up against the freezing cold.

The problem with the heat of summer was that it was compounded by sand. Imagine you're sweating from a searing kind of heat while being constantly doused in powder-fine sand. It blew everywhere and got into absolutely every nook and cranny. You got sand in your food, sand in your eyes, sand in your hair, sand in your bed and even sand in your underpants. No matter what precautions you took, the sand would still find its way past them all.

But if the heat and sand were painful, the flies were a never-ceasing curse. Like the sand, they were everywhere. They swarmed in black clouds around all forms of human habitation in Iraq and made everyone's life a misery. The quantity of flies wasn't a mystery because one of the earliest victims of the war was the sewage system and the streets generally witnessed regular flows of raw sewage, leading to an all-pervading stench and nutrition for legions of flies. The smell of Iraq is something I don't think I'll ever forget, and I still cringe to think of families trying to raise children in that deplorable situation.

Of course, as if the flies, sand, stench and heat weren't

enough to torment us in our security compound, the insurgents were religiously regular with their nightly offerings of mortar, rocket or sniper fire. We lived mostly in tents and so had little or no protection against a direct hit. But the perimeter of the security compound was very well fortified, so the insurgents had no option but to fire straight up into the air and hope they scored a lucky (or unlucky for us) hit. We knew that we would have to be very unlucky to suffer a direct hit but it was something that always played on your mind. It didn't help matters that our compound wasn't too far from where some US troops were based and we often found ourselves in the firing line of attacks directed at them.

11

BAGHDAD

My next tour in Iraq saw me relocate from Basra north to Baghdad. I didn't discover until I was actually back in Iraq that, instead of convoy escort and personnel protection, I would now be strictly assigned to compound security, guarding the massively fortified Ba'ath Party headquarters where part of the trial of former President Saddam Hussein would be conducted. For the next three months I'd be on security duty at a place which was a major target for Sunni insurgents trying to stop the trial of Saddam. Hart had been sub-contracted to provide security on the complex while it was readied for the trial.

The Ba'ath Party complex was huge and was located right on the fringes of the Green Zone in Baghdad, where all the Coalition offices were based as well as what remained of the foreign embassies and UN missions. This was the safest part of the city, but, in Iraqi terms, that didn't mean we weren't targeted by sniper fire or mortar rounds.

We were entirely responsible for perimeter security on the complex while the construction teams readied the cells, the courtroom buildings and the media centre. We would also be responsible for sweeping the building, every morning and evening, to make sure no explosives were left or arms caches hidden away. Our accommodation complex was located just about two hundred metres from the building we were protecting, wedged to the side of a huge US Army compound. Our entire security team would be sleeping in tented accommodation and, the very day we arrived, we erected sandbag security walls around all the tents. We now had some protection from incoming fire to the side but, with only canvas over our heads, we were completely vulnerable to fire coming from above. Directly behind our complex was an area where the US troops stored their Humvee vehicles – which is probably one of the reasons why insurgents close to the complex frequently fired straight into the air.

After a few days of settling in, I realised that the duties weren't too bad and were certainly a lot less dangerous than the convoy missions through insurgent-held areas. The convoy teams were really being hammered now, particularly those that had to run the gauntlet of the Sunni areas north of Baghdad near Fallujah. At the Ba'ath complex, we generally worked twelve-hour shifts, usually seven or ten day-shifts and then the same number of night-shifts. We maintained a tight security regime twenty-four hours a day on the complex, but the major threats came from suicide bombers, snipers and mortar teams who were targeting the US compound behind us. Unfortunately, any rounds that fell wide or short usually hit our complex.

One day I was on security patrol when a huge explosion

shook the ground. I realised that a bomb had gone off behind the massive security walls that protected parts of the Green Zone. The entire sky seemed to darken from the dust thrown up by the explosion. I was far enough away that I was in no danger, except for my ears which were ringing from the blast. When I examined the scene later it was total carnage – the bomber had reduced himself, or possibly herself, to little more than charred meat which was now spattered all over the roadway and up the blast wall. But, yet again, a number of innocent people, trying to go about their daily lives, had been killed for no better reason than being in the wrong place at the wrong time.

Often the insurgent mortar teams would get a lucky hit. One evening we heard the distinctive howl of a mortar round coming in and all of us in the tented complex involuntarily braced ourselves for the blast. Luckily the round went directly over us and struck the US vehicle compound behind us, exploding two Humvees. But, that apart, we knew we had a lot safer detail than the contractors out with the convoys. The problem was that, after a couple of weeks, you tended to suffer from claustrophobia from being in the complex all the time. There was very little to do apart from read, watch TV, play computer games or maybe get a poker school started. A few of the lads would insist on travelling down the road to a US-run gym, but it was something I never particularly liked.

My first tour in Baghdad lasted just over three months and, to be honest, I was glad to get away from the Ba'ath complex. I was beginning to get cabin fever and, despite the dangers, I reckoned that the convoy missions were prefer-able. I came back to Ireland for a short break and, when Hart contacted me about my third tour, I presumed it would be on convoy details. I returned to Baghdad only to be told that, for logistical reasons, I would have to go back to the Ba'ath complex. I had very mixed emotions about it – it was infi-nitely safer than going out on the road, but it felt like provid-ing security for a prison from inside one of the cells.

In the end, this third tour didn't last very long. In fact, in just eight weeks we were informed that the original com-pany that had sub-contracted the Ba'ath project to Hart now wanted it back. The construction teams were almost fin-ished and, I presumed, the company wanted their own con-tractors providing security when the build-up to the actual trial began, for obvious publicity and promotional reasons. I have to say that I wasn't sorry to go, though I knew full well that, within our blast walls and Green Zone security, we'd been shielded from the true horror of what was going on.

Baghdad itself was a nightmare. The city bore all the hall-marks of a place under siege and vividly reminded me of the horror of Sarajevo back in the 1990s. Most buildings bore evidence of shelling and sniper fire. Almost every day was marked by a plume of smoke on the horizon that testified to another car-bomb or suicide attack. And everywhere you looked there were people on edge – from the new Iraqi police to Interior Ministry troops and from US Army patrols to private security contractor convoys. The strongest tell-tale sign of a war zone was the way in which ordinary people moved about the streets – with a wary look in their eyes, ready to dash for cover at a moment's notice.

In the space of two years, Coalition troops had gone from being liberators in Baghdad to targets. The security situation was now so bad and attacks so commonplace that most US and British troops relaxed only when they were back in their heavily fortified security compounds. Body armour was absolutely essential for all security operators either military or private, and some were even investing whatever money they made in state-of-the-art, lightweight armour that would even stops rounds slightly in excess of the 7.62mm calibre. The executions, kidnappings and atrocities had reached such an appalling level that there were days when you wondered whether the entire Government and civil structure was about to implode.

The one good thing about Baghdad was that I was surrounded by old friends. There was a very strong Legion contingent within the security operations in Baghdad and I found myself catching up on old times with lots of old comrades. There was Denis B, who was as good a Legionnaire as ever wore the Kepi. I'd met in Denis in Basra, and, like me, he had been transferred north to Baghdad. There was even Sergeant Akihito Saito, a Japanese national who had been part of my training cadre back at Castel. He was a model soldier, and the kind of guy you liked to have by your side. Before joining the Legion he'd been a member of the elite airborne brigade within the Japanese Self Defence Force. Eventually, he'd had his fill of the new Legion and left, only to drift into private security work.

Iraq had by now attracted security contractors from

almost every corner of the globe. There was Sean L. Sean was in his forties and had been in hot spots from sub-Saharan Africa to the Middle East. He was also gifted at reading maps and being able to plot a course both into and out of insurgent-held areas. Our team leader was Yves M., a former French Army veteran who was in his early forties. Yves had married a Croatian woman and had gone to fight with the Croatian Army during the Balkans wars of the mid-1990s. He was an incredible guy – charismatic, enthusiastic and determined. Like Sean, he also boasted a wealth of experience.

Virtually every security contractor in Iraq knew their business inside out. From weapons to clothing and from surveillance duties to mission planning, these guys were the best of the best. I found the two key issues with clothing in Iraq were comfort and practicality. The scorching heat and humidity meant that cotton was the most comfortable and the clothes we favoured were loose-fitting so that body armour or even straps for our sidearms weren't too obvious. Virtually all of us wore khaki slacks with plenty of pockets for extra rifle magazines, a small canteen and even maps and a compass.

A bit like Cambodia, I couldn't understand how people in general were so friendly and helpful despite the misery and suffering all around them. Once again, it was ordinary people who were paying the bulk of the butcher's bill. I reckoned that for every US or British soldier killed, there must have been at least a hundred Iraqi civilians dying. This was war at its most brutal, most vicious and most merciless level. People were being slaughtered for belonging to the wrong sect of Islam, for living in the wrong area, for having wealthy parents, for being educated, for working for the Coalition and even for being friendly with people who were perceived as enemies. Most of all, people were being slaughtered for the oldest of reasons – for being in the wrong place at the wrong time.

Furthermore, the difference in approach and emphasis between the various European and American PMCs couldn't have been greater. The American security operators always seemed to smack of an operational arrogance. They simply didn't do low-profile, no matter what the circumstances – it was very much in-your-face, shock-and-awe kind of stuff. Don't get me wrong, they had some fantastic people, some really experienced operators, most of whom were drawn from the US military, a lot of ex-Marines and special-ops veterans. But their whole approach seemed to reflect a cultural thing with the Yanks – there was their way and the wrong way. Brute force was always preferred over discretion, and any armed Iraqi was a potential threat. Virtually every US vehicle carried a warning sign on its rear, written in both English and Arabic, that any vehicle which approaches too close will be shot at. That wasn't all. It was often the case that a convoy with US PMCs would open fire when approaching a crossroads or junction just in case any other vehicle might pose a threat. It was very easy to understand how sentiments amongst ordinary Iraqis had swung so violently from viewing foreign troops – both soldiers and security operators – as liberators to suddenly regarding them as occupiers. It was true to say that driving too fast, coming too close to a US truck or simply approaching a junction at the wrong time could get you and your family blown to pieces. Even a few European PMCs came within a hair's breadth of falling victim to 'friendly fire'.

Over time, those of us working with British or European PMCs saw the grim, humorous side of it all. The butt of most of our caustic comments were Blackwater, one of the largest US PMCs in Iraq, whose personnel became renowned for their bullish behaviour and fascination with mirrored-sunglasses and bandanas. Some of the Blackwater guys were okay but others were probably best described as 'red-necks'. Some had never been outside the US before and clearly considered anything that didn't have a US flag stamped on it as suspicious and potentially dangerous – including us.

Most European PMCs were wary when operating in the same vicinity as some of the Blackwater teams because, almost inevitably, there would be contact of some nature. US PMCs were notorious for being willing to shoot at any-thing that approached their convoys at a crossroads or junc-tion, whether it was a European PMC team or simply Iraqi civilians.

Once, at our operations base, a wit posted a note on the medical bulletin board warning of a potential lethal infec-tion known as 'Blackwater Fever'. The symptoms, the note advised, included an obsession with carrying more than one weapon, a desire to wear wrap-around mirrored sunglasses, a red bandana, a crew-cut hairstyle and a goatee beard. Other symptoms included a love-affair with driving around in a powerful jeep adorned with a six-foot-high Stars and Stripes flag while suffering from a persistently itchy rash on the trigger finger. In the context of what was happening in Iraq many of us wondered whether to laugh or to cry at the joke poster.

No matter what security approach was adopted, the issue of life-or-death could be determined by something as small as taking a particular turn at a crossroads. My old training Sergeant in the Legion, Saito, died because of such a twist of fate on 9 May 2005. He was in a security car escorting a convoy of articulated lorries when, on the outskirts of Baghdad, they had to make a decision about whether to turn left or right. The turn they chose brought them in close proximity to a built-up area and, just as the convoy was passing, the insurgents launched their ambush. By now, the attackers knew the strategy of how to attack a convoy – block its exit route, disable the lead vehicles, direct all fire on the security cars and then mop up the rest.

And that is precisely what happened. The security car carrying Saito tried to respond to the attack but, as it stopped to deploy a defensive perimeter, the car was raked with machine-gun fire. Apparently, Saito died like a true Samurai – with his rifle in his hands, fighting to the end. But he never stood a chance. He was killed in a hail of gunfire. Tragically, like so many private security contractors who are killed in Iraq, his body disappeared from the ambush site and has never been located. But all his personal belongings, such as his ID and his employment card, were taken, photographed and the images passed on to Middle Eastern TV networks.

But part of the convoy managed to extricate itself from the ambush and get back to safety in Baghdad. Needless to say, the insurgents gruesomely exploited the killing for its full propaganda value. From their perspective, it was a propaganda bonanza that a Japanese national had been killed and it guaranteed the ambush would receive maximum prime-time coverage. It was big news in Japan because, just twelve months before, an innocent civilian hostage, Shosei Koda, had been beheaded by Abu Musab al-Zarqawi.

This time the insurgents claimed – wrongly – that Saito had been taken alive after being seriously injured in the ambush. The terrorist group, Army of Ansar al-Sunnah, not only claimed credit for the attack but even arranged for multiple photos of the dead Sergeant's body to be posted on the Internet. In Tokyo, the best-selling national daily, *Yomiuri*, carried a lengthy report in which a Government official spoke of how difficult it was to establish communication channels with such a shadowy terrorist group. But my friend Saito was now beyond worrying about such negotiations. He'd died a true soldier's death – fighting bravely to protect those in his care.

At this time I was back in Ireland enjoying a break from my abbreviated tour at the Ba'ath complex. I'd only been home a couple of weeks when I got a call from the Hart Group in London. There was a contract available for Abu Ghraib, a woman with a clipped Home Counties accent explained. Was I interested? I paused and then laughed at her. Was she serious? I had been in Iraq for over a year and knew that some contractors regarded working out of Abu Ghraib as akin to a death sentence. 'Well, if you're not interested I suppose I will have to call someone else,' she said. I think it was the tone of her voice that flipped a switch in my head and, before I had time to analyse the pros and cons, I'd accepted the contract and was told I would be en route to Baghdad within four days.

To be honest, a part of me was also concerned that, if I refused this contract, another job might not be offered to me for some time and, despite the dangers, working for Hart had reinvigorated me. I felt as good as I had back when I'd been determined to make the Legion my life. In fact, working for Hart was a bit like being back in the Legion again, only this company was run infinitely better than the Legion. However, I went for a pint that night in the local Roaring Donkey pub and bumped into a friend, Brock Carlos. He asked me what I was up to and I told him about the Baghdad convoy contract. He bluntly warned me he thought I was mad to go back given all that he'd seen about Iraq and Baghdad on the TV news. I admitted to Brock that I had a bad feeling about this trip – for the first time ever – but I was going to go anyway.

A few days later I was on a plane bound for Kuwait. I transferred through the airport and caught a Royal Air Force transport plane to Iraq. As I walked through the Baghdad airport processing area, I happened to run into two Hart officials that I'd come to know. Both were decent guys and we stopped for a quick chat. In friendly fashion, they asked what I was doing. 'Where are you heading, Paddy?' one smiled. I simply answered 'Abu Ghraib', and instantly their smiles faded as they traded a look. They tried to ease the atmosphere by cracking a few jokes about the type of lunatic contractors that looked for work in Abu Ghraib, but their initial reaction spoke volumes about how bad it was going to be. The security situation in Iraq was spiralling out of control and my new contract had placed me right smack in the middle of the fire-storm. It was by far the most dangerous work I'd ever undertaken in Iraq and, as I arrived 'in country', I heard on the grapevine that convoys from Abu Ghraib were being hit on an almost daily basis. Just to make matters worse, the security situation in Baghdad was deteriorating so fast that there were now even parts of the city where US patrols would transit only if they were in force and with heavily-armoured back-up on standby.

I found it hard to believe it was the same country I had arrived in just one year before. Everywhere you went, you could almost slice the tension in the air. The two Islamic sects – the Sunnis and Shi'ites – were now at each other's throats and, to be honest, I felt that a civil war had already started. Attacks on mosques were now becoming commonplace and it was clear that the insurgents were trying to incite the two sects to slaughter each other with the Coalition forces caught in the middle. Wedged alongside the Coalition forces were us civilian contractors and we were beginning to be killed in numbers matching those of US troops.

It was now so bad that most US troop movements were simply going from their fortified bases to their objective, and then straight back to base. The countryside in between was akin to a World War I no-man's land. Furthermore, Iraqi civilians did not appreciate any efforts to contact them publicly. Simply being seen to deal with a US or British solider in public, even for something as small as selling cigarettes or fruit, was enough to earn the attention of the insurgents or the death-squads. The sectarian violence was so bad that tens of thousands of Shia and Sunni families were now fleeing the neighbourhoods that they'd called home for generations and, for safety, were congregating in areas dominated by their own sect.

The death squads were roaming the city at will and the rumours abounded that Iraqi police and Army units were taking an active role in the sectarian bloodbath. Even more sickening was the fact that the bodies of hundreds of the victims showed evidence of horrific torture before the poor unfortunate was finally shot or beheaded. Each morning, fresh bodies could be found by the roadside, in fields or even dumped by marketplaces. Baghdad was fast becoming a charnel house. And this would be my home for the next three months.

And yet, despite the dangers, the Iraqi drivers and security contractors continued to show up for work with us. I suppose they really didn't have a choice – if they didn't work, their families would go hungry. I have enormous admiration for these guys. They were doing a thankless task in horrific conditions with huge implications for their own safety. Yet they were always friendly, co-operative and did the job to the best of their ability. God knows, they deserved better than to see their country being submerged in a sea of blood.

Looking back, if I had had any sense, I would have quit Abu Ghraib and looked for another contract. It was dangerous to the point of being lethal. One of the British contractors grinned at me one day and said: 'Paddy, if you want to prove you've got the nuts, you've come to the right place.' Our work solely involved convoy escort – or, as one colleague grimly admitted, driving around Baghdad as 'armed bait and ambush fodder'. I didn't know it at the time, but I was effectively the replacement for my old friend, Sergeant Saito, who'd been killed only a few weeks earlier.

After settling into Abu Ghraib and learning that I was effectively Saito's replacement, an old nightmare flashed back through my mind: the 'rule of three'. In Bosnia, I'd cheated injury when my buddies Fotyi and Houlsek had been hurt. Now here I was in the most dangerous part of the most dangerous country on earth and two of my friends had already been killed, Nick Pears and Akihito Saito. Would I be the third or would I cheat the fates once again? It certainly didn't make for relaxing thoughts inside the security compound at night.

The one bonus was that, because of the nature of the security situation around Abu Ghraib, only the most experienced and capable operators were on staff. Anyone who was on contract duty here knew their stuff, and the whole operation was run by an ex-Royal Marine commando named Geoff. He was an incredible operator and a really decent bloke as well. His job was thankless: he had to try and provide security for convoys that were being sub-jected to an onslaught of attacks, but he also had to try and keep both his Western and Iraqi security contractors happy despite the fact they were being hammered on a daily basis by insurgents.

Mind you, some routes were a lot better than others. Deliveries around Baghdad city itself weren't too bad, par-ticularly if you could use speed to get in and out of the area fast. The one place no-one wanted to go near was Fallujah, referred to as 'Insurgent City'. The area was known as the Sunni Triangle, and the part wedged between Fallujah and al-Habaniyah was literally a killing zone. Even heavily armoured US Army patrols were wary of moving into this area without air cover.

But the personnel made the duties bearable. Yves was a wonderful character and, I guessed, had been the kind of officer that every soldier prays he will serve under. Yves led by example and he never asked anyone to do work he wouldn't do himself. Being French, he had a great appre-ciation of the Legion and we hit it off straight away. Yves was very friendly with Sean, an ex-South African Army soldier. Both men were middle-aged, had families back home and didn't take any unnecessary risks. They also complemented each other in terms of their skills – Yves had a wonderful rapport with the security contractors and, in particular, the Iraqi personnel, who were devoted to Yves because he drew no distinction between them and the Western security contractors. The Iraqi security operators undertook work for Yves that they would have refused to do for anyone else. And with Sean's genius at plotting routes and memorising maps and terrain, I knew that if he was with our security team we would never get lost, a hugely important issue in the maze of roads, alleys and laneways around Baghdad.

But the major bonus for me was that I'd be working alongside my old friend, Denis B., in Abu Ghraib. Denis was a natural soldier and as loyal a comrade as anyone could ask for. He'd spent almost thirteen years in the Legion's commando unit and it was only in Iraq that we discovered we'd both served on the Drushina Valley mission in Bosnia all those years before. Off-duty, I spent most of my time with Denis and, even on convoy duty, whenever we got the chance we paired off and enjoyed a few hurried cigarettes. I suppose it was typical of the tension in Abu Ghraib that, while I normally smoke about twenty cigarettes a day, in Baghdad I was burning through between forty or fifty a day. Denis was the same.

I had expected the security situation to be bad, but, in truth, it was a total nightmare. One time we had to take a convoy north of Baghdad towards Kurdistan and by the time we arrived back at Abu Ghraib we found eight IEDs had been planted around our perimeter. Attacks on convoys were now assuming ferocious proportions and the insurgents, far from being intimidated by the proximity of Coalition military patrols, were now willing to engage them in running fire-fights. US and British losses were beginning to soar and new and more lethal IEDs were starting to wreak a terrible toll.

One day, I was in an escort car when, just as I drove

alongside one of our artics, it had a tyre blow-out. The noise was incredible, but the Iraqi driver kept control of the truck and slewed it over to the side of the road. The other escort cars swept around and set up a defensive perimeter. But I had to sit in the car – I had been convinced the noise was an IED going off and was ready to cut loose with my AK. I sat in the car and practised slow, deep breathing for about five minutes before my heart-rate came back to normal and my gorge settled back down in my throat. The other lads knew precisely how I felt. But that was life in Iraq – lived on the edge, waiting for disaster at any second.

Finally, the bad news came. Geoff told us on the morning of Friday, 4 June that we'd been assigned to a new work order. We all knew that meant convoy escort – but to where? Geoff confirmed our worst fears when he said that the convoys were bound for al-Habaniyah and would be undertaken on a daily basis. Geoff said that Yves's team would take the convoys for the first two days – Mondays and Tuesdays – and the second team would operate on Wednesdays and Thursdays. Personally, I thought that was better for us because, by Wednesday, I reckoned, the insurgents would be more than ready to hammer anything that travelled that road.

And then it all started, slowly but inexorably, to go wrong. Our security team worked really well together. We all knew and respected each other and by now had formed a seamless unit. But Sean had been caught in a pretty bad ambush just two weeks before and Yves was due to finish his contract and go on leave in three weeks' time. Security duties to al-

Habaniyah were hardly the way to prepare for life outside Iraq. Worst of all, our Iraqi contractors threw a fit when they heard about the convoy routings. It was hard to blame them. Even the dogs on the street knew that, for many convoys, a trip to al-Habaniyah or Fallujah was a one-way ticket.

In the two days before the first trip, even our ordinary convoy escort work began to suffer from problems. Driving through Baghdad city, our convoy was badly separated by cars and we didn't know if they were insurgents or simply gangsters looking for an easy target. Our security team radios were humming that day as we shouted at each other to try and co-ordinate a response and keep the trucks together. The whole day was a disaster and we were exhausted from the combination of the relentless traffic and trying to keep the convoy intact, while not wanting to open fire without good reason.

At times I felt like a sheepdog, except that I was herding old artics. One of the tactics we had evolved was to have the trucks travel at different depths and in different lanes. If we were spread across a four-lane highway in heavy traffic it was harder for insurgents to spot that it was an actual convoy. But that increased the workload on the security team by a tremendous amount. We were constantly speeding across lanes from one truck to another to make sure everything was okay and, because of the level of separation, we had to be very vigilant to possible insurgent vehicles wedging themselves beside a vulnerable truck.

All the hassle didn't help our frame of mind. As we approached the departure date for the first convoy to al-

Habaniyah we already felt like we'd been put through the mill. Our tiredness wasn't helped by the fact that the convoys from Basra and the south timed their runs through the late night and early hours of the morning which meant they arrived at Abu-Ghraib in a bellow of screaming engines and security team shouts from 5.00am onwards. No-one within a four mile radius could have slept through that din.

And finally the dreaded day arrived. When they heard the destination, our Iraqi contractors refused to go. Every single one of our fifteen Iraqi security men quit on the spot – they said they wanted to keep their jobs but valued their lives more than a trip to al-Habaniyah. Privately, I was thrilled. If the Iraqi lads wouldn't work, the convoy simply couldn't go and we could all relax. I found a quiet spot with Denis and began chain-smoking, all the while half-praying that the Iraqis would remain stubborn in the face of all the enticements offered to them. Finally, Geoff arrived at the scene and he and Yves began talking to the Iraqi security lads and, gradually, I could sense that a few of them were slowly coming around after appeals to their sense of duty and courage.

But the rest of us were nervous too. From my perspective, the choice was clear: I could bow to my inner anxieties and walk away from the job, or I could stand by my mates and do what I had been trained to do. For me, it didn't even rank as a choice; there was no way I'd walk away from my mates. I have to respect the way Geoff handled the situation. He didn't scream or shout at us the way some officers might. He didn't tell anyone they were a bastard for refusing to escort the convoy. Typical pro, he simply shrugged his shoulders and said: 'Lads, the convoy has to go, so who's going to take it?' He was basically addressing the pride of the contractors and there were some very proud soldiers in that compound. I could see it from his point of view. We were there to do a job we were being paid to do that job. So what was the problem? We all knew the risks when we signed up.

The convoy required four Western security officials and fifteen Iraqis. But now eight of the Iraqis were willing to go though the other seven still adamantly refused to budge. After a few more minutes of negotiations, Geoff and Yves called Denis and myself out to the compound. 'Fill in the blanks, Paddy,' Geoff ordered. So I went down to our accommodation block and directed the first seven Iraqi security guys I met to get their kit together and join the convoy. Most of them realised that if they didn't travel with our convoy today, they'd be on security detail tomorrow. It wasn't much of a choice.

So we re-organised the security teams and got our escort cars ready. I'd be travelling in the back of an old BMW. My driver would be a young Iraqi called Arkan and my Iraqi security contractor would be Wisam. Arkan had only been driving with me for the previous ten days, after my previous driver, Sam, was re-assigned. Sam was a superb driver and could read a situation on the road in seconds. Arkan was a nice lad, but didn't have Sam's experience.

On one of my first security details with Arkan, I was scanning the road behind when suddenly I heard the blast of gunfire at close range. I swivelled to see Arkan driving the car with one hand and blasting away with the 9mm out through the front passenger window. 'What the fuck are you shooting at? Where, where?' I shouted as I levelled my AK in the direction Arkan was firing. The young Iraqi excitedly explained that a car had got too close to him on the highway and he immediately assumed it was an insurgent. I told him to calm down and put away the gun. We never saw the other car again and, to this day, I don't know whether it was an insurgent attack or just an overly aggressive motorist. But in Iraq no-one took chances on the roads and if you approached another car at speed and in an aggressive manner, you could expect a response. I'd never worked with Wisam before, but I knew he'd been with the Abu Ghraib team for a while, so he knew his stuff. He would take the front-seat position alongside Arkan, leaving me the entire rear seat for my kit, body armour and spare ammunition.

Wisam was armed with an AK-47, but Arkan, as the driver, wasn't routinely issued a firearm. But I never liked to have anyone in my car unarmed, particularly given the security risks we faced on a trip like this. I'd given away the Makarov I'd bought in Basra and had replaced it with a Tariq 9mm, a cheap, licensed copy of the Italian Beretta. Somewhat ironically, Saddam Hussein had bought the rights to the Italian gun after the conclusion of the Iran–Iraq war, apparently unhappy with the Soviet pistols on which his army had been relying. The Tariq was almost identical to the Brazilian Beretta clone, the Taurus, and while the Iraqi gun was cheap and basic, it was also lethally effective at close range and boasted a handy magazine capacity of fifteen rounds. I handed Arkan my Tariq, making sure the safety catch was on, and told him to keep it by his side at all times. Given where we were going, I figured he would probably need it.

One precaution I insisted on was that we carried as much spare ammunition for the Kalashnikovs as we possibly could. I reckoned that if we were deliberately going into harm's way, we wouldn't want to do it light on rifle rounds.

After checking all the security cars and verifying our radios and contact frequencies, we were ready to roll. Every single member of that security detail knew that there was likely to be trouble ahead. But we had a good leader, we had the best training available and we knew that those civvy lorry drivers needed our protection. We also knew that if we refused to do this run, some other security team would have to take up our slack. It was that simple. And, ultimately, our professional pride won out. We had all been part of elite military units in France, Croatia and South Africa and they all taught that fear was something you conquered, you just didn't run away from it.

Geoff wished us well and Yves ordered the five cars out of the compound. I was assigned to the 'CAT' or counter-attack-vehicle, so it would be my job to keep the convoy together, scout for trouble and, if necessary, lead the fight-back to allow the others time to get clear. I don't know what switch flicked in my head but, once the formal decision to go was made, I was perfectly calm. I knew a lot of the lads were worried at the prospect of what lay down the road but I was totally focused on my job and my training. If every-one did their job properly, well, at least we'd have a

fighting chance. We slowly made our way down the fortified approach route to the Abu Ghraib complex and, leaving
the safety of the security zone, turned out on to the highway
bound north for Fallujah and al-Habaniyah – right into the
arms of Iblis, Islam's own devil.

12

THE KILLING GROUND

It was the perfect killing ground. The insurgents hit our convoy at precisely its most vulnerable point – half-way through negotiating a major turn on an exposed, elevated road midway between al-Habaniyah and Fallujah. The raised roadway left us open to fire from all sides and the turn in the road meant our heavy trucks had to slow down to a crawl. We were literally in the back-garden of Iraq's insurgency – and, indisputably, in the most dangerous place on earth. I knew that the next few seconds would be crucial in determining our fate. If we could keep our lorries and security vehicles rolling and get clear of the main zone of concentrated fire we might have a fighting chance of getting out.

Arkan had accelerated to get us away from the turn and had managed to push our Toyota in under one of our trucks, which offered us some protection. We had already lost Wisam and I had been injured, but the incoming fire was coming so fast the pain didn't register. I had combat-crawled to the protection of the Toyota. From behind us all hell broke loose.

As the rounds slammed into my Toyota I knew that the ambush site had been chosen with precision and deadly cunning. I also grasped that, if I had immediately ordered Arkan to drive on, we would have had a great chance at escape. We could manoeuvre around Yves's wrecked car whereas the convoy trucks simply could not. In seconds we would have been clear. But running away is not something the Legion taught me. I was here to protect the convoy, defend the truck drivers or die trying. These Iraqi drivers had put their lives and trust in our security team and I simply couldn't run away. I wouldn't have been able to live with myself afterwards. Even more importantly, my friends were already fighting for their lives and needed every bit of defensive firepower available.

The insurgents had cleverly waited until all our security vehicles had stopped around our five stranded front trucks before they unleashed their main weapon, a heavy Russian-made PK machine-gun set up on the roof of a two-storey building to our left. It was a squad-type weapon which boasted a terrifying rate of fire. The instant that machine-gun opened up, it began to cut our convoy to pieces. Any chance we had of fighting our way through the ambush disappeared the second that Russian gun opened up on the stranded vehicles and wrecked them, one by one. We were now taking heavy fire from three sides and only our left rear flank, leading from the elevated roadway in the direction of marshy fields stretching towards al-Habaniyah, seemed clear of insurgents.

Yves, our charismatic team leader, immediately led the

fight back. The initial salvo of gunfire from the AK-47s had disabled his black Toyota car and, as it slewed across the road, it blocked the five lead convoy trucks. Yves got out of the vehicle to return fire. The other security car, carrying four Iraqi security guards, was also raked with AK rounds. One round struck the car's engine-block and it slewed, crippled, across the road. But, worst of all, the two crippled security cars had combined to force the trucks following immediately behind to slow down and stop, creating a fatal log-jam. Now their cabs were being sprayed by gunfire from insurgents hidden in the houses and trees to our right. The civilian drivers were either dead or dying. Those who got clear where hiding underneath their lorries and scrambling desperately for cover from the thundering fire.

The scout vehicle, a blue Opel saloon, carrying four Iraqi security contractors, was initially the focus of the insurgents' gunfire and at least two of the four guards were killed instantly. The insurgents had wanted to cripple this lead car and they had succeeded admirably. The engine was torn apart by incoming rounds, leaving the car little more than a hulk. The other two contractors desperately tried to get back towards Yves's car and form a temporary defensive perimeter.

Sean was now our main hope because his Opel estate was carrying our only heavy machine-gun, a Soviet-era PK. We had to get that gun into action fast – it just might give us enough firepower to keep the insurgents' heads down and force them back from the convoy until help arrived. When the ambush erupted, Sean's car was towards the rear third

What else can you do except smile in the face of danger? Despite my happy exterior I was deeply worried about the security situation in Iraq, and knew that even my trusty Kalashnikov might not be enough to protect me.

The security team Charlie Three One pose with our Iraqi drivers when we arrived safely in Kurdistan after a high-risk convoy mission. The Kurds treated us like visiting royalty. I'm third from the left in the front row; Sean L. is standing on the far right; Wisam, third from right in the front row; Doc is on the far right of the front row; Heider is sixth from left in the back row; Ahmad fifth from right in the back row – he was the one from whose dying body I had to strip ammunition during the al-Habaniyah ambush; fourth from the right in the back row is Big Ali, who was one of the last to die by my side in the ambush – it was his rifle-burst that almost shattered my eardrums.

Left to right: Sean, Yves, Dennis – taking a break but with their AK-47s resting on the table in front of them.

My old Legion sergeant, Akihito Saito, points to the tell-tale smoke plume of a bomb blast in th distance. A veteran of Japan's elite airborne brigade, Saito was killed in an ambush just a few weeks before I nearly died at al-Habaniyah. Note the nondescript yet powerful saloon cars bei used by the security team.

TOP TO BOTTOM:

The horrific aftermath of an insurgent attack.

A typical Private Security Detail vehicle interior after an ambush.

Security advice, in both English and Arabic, that you ignored at your peril. Every US Army Humvee carried this type of warning on its rear bumper – and any vehicle, be it driven by an Iraqi family or Western security contractors, would be shot at if it didn't comply with the 'stay away' advice. Sometimes, drivers risked being shot at if they simply approached a crossroads at the same time as a convoy.

CAUTION STAY 100 METERS BACK OR YOU WILL BE SHOT

تحذير ابقى بعيدا المسافة ١٠٠متر
او سوف تتعرض للاطلاق النار

A convoy truck burns fiercely after another insurgent ambush. Their attack took place on the road to Nasariyah, one of the insurgents' favoured hunting grounds.

The stomach-churning reality of war in Iraq.

A Sikorski Chinook flares for landing as a British Army team sets up a snap roadside checkpoint. The Chinook, along with the Blackhawk and Apache, have become the symbols of the war in Iraq. After the ambush at al-Habaniyah I was evacuated to hospital in a US Chinook.

An insurgent lies dead by the roadside with a US soldier standing over him after an ambush on a convoy backfired. The convoy security team, with assistance from a US military patrol, fought back and the insurgents ran when several were killed in the fight. The insurgent's bottle of water stands mutely beside him as well as two unused Russian-made RPG-7 rockets.

A good example of the type of security that was evident on most Baghdad streets and yet which failed to curtail the tide of violence. Any vehicle that approached such a checkpoint at speed would be fired upon.

The view from the top floor of the Ba'ath Party headquarters in Baghdad where Saddam Hussein once wielded his terrifying power, and, somewhat ironically, was later put on trial for crimes against humanity. The tents surrounded by sandbags are the accommodation blocks used by our security team. Note the heavily reinforced blast wall to the right of the photo. This is Baghdad's Green Zone, and, directly behind our tents, lies a major US Army compound that was frequently targeted by insurgents.

The chill of a winter night in Baghdad is evident from my warm clothes. This was the relative safety of the Green Zone and yet the barbed wire and blast walls tell their own tale.

Travelling in convoy on an Iraqi road – in this situation you were always on the alert for threats from other vehicles and constantly crossing lanes to confuse attackers.

A general street scene in Haiti showing the dirt and the squalor visible everywhere.

A Haitian voodoo dancer adds a chilling touch to a street carnival scene.

of the convoy. The rear or 'back door' of the convoy was being held by Denis and his Iraqi team. I didn't know it then, but Denis's old Mercedes saloon was trapped directly on the apex of the turn when the shooting started. Now, Sean's Opel flashed towards the front of the convoy and then abruptly skidded to a halt after a bullet tore through its engine block just metres from Yves's already crippled vehicle. Now we had four of our five security cars disabled and the fifth, my old white Toyota, was quickly being torn apart before my eyes.

Sean jumped out of his Opel and began returning fire with his AK, moving to the rear of his vehicle in a desperate bid to get the heavy machine-gun into action. But the odds were simply too great. I can only guess, but there must have been fifty to sixty insurgents surrounding our convoy and their fire was withering. They were hidden in a maze of roadside buildings and it was difficult to direct aimed fire at them. Any movement at all by our team attracted a hail of fire, and while the insurgents weren't known for the accuracy of their shooting, there were so many AK-47s in action here that it didn't really matter. It was like shooting at a barn door with a shotgun – it was virtually impossible to miss. The situation was moving fast from bad to desperate. The only small mercy was that they were using the same weapons as we had, and there was no sign of any rocket propelled grenades (RPGs) which, given our situation, would have been absolutely disastrous for us.

I think the full realisation of the horror that faced us only dawned on me when I saw Yves's and Sean's cars being peppered with rounds. I'd already seen Wisam killed. But the full force of what we now faced vividly hit home when I saw Sean's driver, Tamir, take a round and then stagger out on to the open roadway. His body immediately began to jerk and shudder as repeated insurgent rounds struck him. Even when he fell, his dead body still jerked from the impact of incoming rounds. The poor bastard was clearly dead, but the insurgents were still taking pot-shots at him. Grimly, I realised that this was now a fight to the death. 'We are in serious shit here,' I muttered to myself.

Sean and Yves were now, with what was left of their Iraqi security teams, huddled behind their cars and desperately returning fire at the various insurgent positions. There was a gap of ten to fifteen metres between their two cars and another similar gap back to the two convoy trucks between which I established my firing position. We were badly exposed and I knew it. We had to consolidate our position or we risked being over-run individually in small groups. Sean and his team had no choice now – they had to brave the fire and make it back to where my Toyota was wedged underneath a truck. If they stayed where they were, they were certain to die because of the lethal cross-fire directed at them.

I was trapped in a confined area between my wrecked saloon and two of the rearmost convoy artics. But at least I had substantially more shelter than either Yves or Sean. We were taking heavy fire from both the right and left side of the road and the insurgents' heavy machine-gun was chewing up the vehicles one by one. We all kept very low, in a semi-crouched stance, because we could hear the constant 'phizz' of AK rounds over our heads. Worst of all, the elevated position of the road telegraphed our every movement to the insurgents. The only stroke of good luck I had was that the civilian truck which had been passing when the ambush erupted was now offering us an element of cover from the machine-gun. I guessed that the heavy cab and engine block of that truck reduced the arc of fire of the machine-gun by about fifty percent, a stroke of good fortune that I realised had kept us alive until now.

I spotted one insurgent running down by the berm, below the level of the road surface, and realised he was trying to out-flank Yves's position. I swung my AK around and fired a short burst, but cursed when I saw the rounds miss, high and wide. The rear doors on Yves's car were still open and the right rear door blocked my view of where the insurgent had gone. Without hesitating, I fired another burst through the door and its window, trying to clear my field of fire to track and kill the insurgent. Yves swung around and stared at me as much as if to say: Aren't there enough of these fuckers shooting at me; why do you have to take a few pot shots too? Then he grinned and winked at me, which was typical of the guy. We were fighting for our lives and yet he was still as cool as a lump of Arctic ice. He was one of the most talented natural soldiers I'd ever served with and men would have followed Yves through the gates of hell. Unfortunately, that is precisely where we all now found ourselves.

By this stage, all the drivers who hadn't been killed or wounded were now huddled together in the narrow space between my bullet-ridden Toyota and the convoy trucks.

Arkan was using my Tariq 9mm and was shooting at the insurgents with gusto, though I doubted if he was coming anywhere near any of them. When Sean and Yves finally made it back to my position, I estimated that there must have been about sixteen of us left, between security personnel and civilian drivers. But one Iraqi driver, whom we had nicknamed 'Doc', found the ordeal simply too much to handle. He had squatted down beside the wheel-arch of the truck, wedging himself into the opening and was slowly rocking back and forth with his hands held over his ears and his eyes bulging. The poor guy looked almost catatonic. I got no response at all from him when I shouted to ask was he okay? The next time I looked, he was slumped over, dead, having taken several rounds.

I was trying to return aimed fire, but it was hard because of the number of insurgents and the mobility they enjoyed. It was also difficult because I now realised I'd been shot in the right elbow. It dawned on me that, when our Toyota first skidded to a halt and Wisam was killed, the rear passenger window had been shattered by several incoming rounds. One of these must have hit me just as I was trying to level my AK to return fire. The combination of shock and adrenalin had left me unaware of the wound until now. But my arm now felt heavy and leaden, yet even though it was beginning to throb painfully I could still partially use it.

Once Sean joined my group, he assumed my firing position and I moved forward, crouching between the cab and fifth wheel of the civilian truck. This allowed me to fire both in the direction of the heavy machine-gun position and towards the main insurgent position to our right and rear. By now, my fatigues were covered in blood, which was pouring from my elbow wound and from my bare arms, which had been shredded as I combat-crawled along the glass-covered roadway. Sean threw forward my body armour and then opened up with his AK towards the insurgents still attempting to outflank us.

If we could maintain a reasonable field of fire until relief arrived, we could hold the bastards back. But the insurgents had smelt blood and apparently decided they could overwhelm our position. They were focusing an awesome amount of firepower on us and the rounds kept whizzing overhead and clanging into our vehicles. We were now on the receiving end of one of the Kalashnikov's attributes: its old 7.62mm round had tremendous 'punch' and, sooner or later, the rounds would start pouring through the scant cover offered by the car and trucks.

About five minutes after I had assumed the new position I glanced over my shoulder to check on the group – only to see Sean die. The South African had been standing by the side of the truck, returning fire, when he suddenly did a half-pirouette and, without a word, fell face down on the ground. As I stared in horror, he was absolutely motionless. Sean was wearing full body armour so I could only guess that he'd been hit by an armour-piercing round. He died instantly. We had just lost one of our best operators.

Yves was now screaming over the radio back to base to get us reinforcements and get them to us fast. The fire from the insurgents was still not easing up and, with their numbers, it was only a matter of time before they caught us in a new cross-fire. An even more nightmarish scenario was that they would slowly infiltrate the convoy itself, killing us one by one as they swarmed around our vehicles and compromised our lines. Yves kept shouting for reinforcements and asking that any American patrol in the area be immediately notified of our plight. I looked at the faces of the Iraqis around me and realised that they all believed they were going to die. But I wasn't about to let anyone give up hope – not yet.

'Choppers are coming, choppers are coming,' I lied to the huddled Iraqi group. A few managed to smile in hope but the constant 'zing' of insurgents' bullets whipping by spoke volumes about the reality of what we really faced. At one point, after I emptied an entire magazine at one insurgent position, I almost convinced myself that I could hear the clatter of approaching helicopter rotors in the distance. But they never came.

Every time I looked around, our group seemed to be getting smaller. There were bodies piling up everywhere now and pools of blood were congealing on the road surface around us. The Iraqi drivers were huddled in our midst, praying for rescue. What shocked me most was the sheer volume of fire we were taking. There must have been between thirty or forty insurgents dug in around us, and they showed no sign of easing back on their attack. They weren't trying to conserve ammunition either, which meant that this killing zone had been prepared with infinite care. There was simply no place to take safe shelter. The rounds were now coming through vehicle doors, car boots and even underneath the crippled artic. I was in the most advanced position, between the truck cab and its fifth wheel, but I reckoned that, ironically, the giant engine block of the artic gave me the best protection we had.

I realised that the US outpost we had passed was only about four miles away. Surely they couldn't miss the fire-fight that was now taking place on their horizon? The explosions and flames must have been visible from ten miles away. I said a quiet prayer that the US lieutenant had already called for heavy reinforcements. But I just didn't know whether we could now hold out long enough for reinforcements to reach us. As I checked the group once again, I realised with a start that Arkan was dead. The poor guy had been fighting back with the pistol which, in our plight, was brave but completely futile. He had been a wonderful bloke, good humoured and willing to learn. I didn't see him fall, but I can only presume that, as the insurgents manoeuvred around us, he was left exposed and was cut down by a well-aimed AK volley. Now, he was just another corpse on Iraq's spiralling butcher's bill.

The slaughter was unbelievable. There were literally piles of bodies surrounding me – men lying motionless on top of each other, each in the precise position they had fallen after being shot. But these were friends, colleagues, comrades. Just a few hours ago, some of these men were laughing or sharing a cigarette with me. Now, they lay sprawled together around me in death's embrace.

I knew we couldn't go back – even if we had a vehicle that could still move – because to do so would have meant trying to cross the field of fire of the PK machine-gun on the roof. We couldn't go forward either because Yves's car blocked the road. Denis, who was protecting the convoy rear, was also pinned down by heavy fire and couldn't reach us. His car was wrecked and I guessed his only hope was to withdraw, fighting, back down the road we had just travelled and look for reinforcements. But I was in the heart of the kill-zone and, sooner or later, the insurgents would move in around us. As the minutes ticked by, it slowly dawned on me that we would soon run out of ammunition if we were forced to maintain our current level of engagement. We had already shot off a huge quantity and I guessed that not even a regular US Army patrol would have had the reserve ammunition to cope with this type of sustained, relentless fire-fight. Subconsciously I decided to try and conserve ammunition and fire only short, well-aimed bursts at identifiable targets. But the danger of doing this was that it signalled to the insurgents that we running low on rounds.

The insurgents kept slowly tightening the net around us and it seemed there was nothing we could do about it. I reckoned that between eight and twelve insurgents must have been killed in our return-fire, but they still weren't backing down. I could hear heavy fire coming from behind us, by the bend where I had first spotted the insurgents. I realised that Denis was also fighting for his life, and that there was no way he was going to be able to make it forward to our position.

I spotted a car coming out of an alley to the rear of Denis's position and move forward as if to block the road behind us. Without hesitation I took careful aim and fired a burst into the cab of the car. I realised that Denis and his team had also opened fire on the car. It suddenly veered away from its original direction and ploughed off the elevated roadway. At least the route was still clear now if Denis and his guys could make a break for it, I thought.

Later, I learned that Denis and what was left of his team managed to hi-jack a pick-up truck and race back towards the US position. When they arrived, their pick-up was like a sieve with bullet-holes – their Iraqi driver had managed to steer the pick-up while lying flat on the cab floor! It was just as well because, when they reached US lines, there were four concentric bullet holes sent right through the windscreen by a sniper at the point where his head would have been. While we didn't know it, the insurgents had now split their ambush force with roughly one-third going after Denis and his team and two-thirds moving in for the kill on my position and the main convoy body.

From my niche I saw three shrouded figures crossing the road behind Denis's old position. I knew they weren't part of Denis's team so I didn't hesitate. I took careful aim with the AK and let blast a volley at the group. I remember slowly saying the 'Our Father' prayer as I fired the rounds into the group, hoping all the while that Denis could make it out of the kill-zone. Even now, the instant I hear that prayer it brings back memories of the ambush. Deep down, I was worried that we didn't have much time left. I had barely finished firing at the three shrouded figures when I felt a searing pain on the left side of my face. The force of the impact drove my head against the side of the truck I was sheltering

beside. Initially, I went numb and the pain didn't fully register. But then I realised that blood was pouring from my face
and cheek. A round fired from the insurgent positions
behind me had creased my cheek. Just one centimetre further to the right and it would have killed me instantly, passing through my brain. One Iraqi security contractor, Ali,
crawled forward and tried to apply a field dressing to the
wound but, because we were in constant action, the dressing
soon began to flap loose.

By now, the fire-fight had been raging for more than an
hour and there was still no sign of help. Our little group,
which had once numbered around sixteen, was now down
to a handful of survivors. Yves was still screaming down
the radio for help but we all knew we were running out of
time. Yves put down the radio, left the channel open and
looked briefly at Sean's fallen body before he levelled his
AK and emptied a full magazine towards the insurgent
positions to our right. Our security control centre back in
Abu Ghraib would be able to follow the full tragedy of
what was happening now. I felt a rising tide of anger well
up inside me as I saw guys I liked and worked alongside
being killed and wounded. I knew that Yves was now
fighting mad and, like me, was determined to make it a
costly last stand for the ambushers.

I also knew that Yves had spotted the open flank along
which we could have tried to make a run for it. Once down
off the elevated road, you could have disappeared into a
large field full of reeds and marsh grass. It was perfect cover
for evading pursuit. But we now had wounded amongst us

and the insurgents didn't take prisoners. Without openly admitting it, both Yves and I knew that if we had made a run for it, the whole ethos of PMCs in Iraq would have been destroyed. What Iraqi civilian would ever entrust his safety again to a PMC security contractor if word got out that two guys charged with protecting the sheep from the wolves had made a run for it just to save their own hides? No, we had set up our position and we'd now have to defend it while we still had ammunition left.

Incredibly, amidst all this carnage, one of the Iraqi drivers sheltering in a truck to the rear of the convoy, finally decided to make a break for safety. He managed to start the truck and, despite the fact that most of its tyres had already been shot out, tried to drive up the road and pull clear of the killing zone. The poor guy never had a chance. As I watched in amazement, he didn't make it very far. The instant the truck began to roll it attracted virtually every bit of fire from the insurgents, including the attention of the lethal PK machine-gun. The truck was torn apart and, as it slowly veered off the road and plunged down into the marshy field to my right, it exploded in a ball of smoke and flames. The driver had obviously been killed in the cab.

Anyone could see that we were now moving into the endgame of the tragedy. The carnage around us left no-one in any doubt as to what was going to happen next. But the only thought that was dominant in my mind was the conviction that I wouldn't allow myself to be captured and paraded in any orange jumpsuit. Whatever happens, Pádraig boy, I silently promised myself, it ain't going to be that.

Our ammunition situation was now critical. I was down to less than two full magazines for my AK and had no option but to start scavenging rounds from the dead and wounded lying around me. Ahmad, who was the heavy machine-gun operator in Sean's car, was lying in a pool of blood on the roadside close by the wrecked Opel. He had taken several rounds through the upper body. But he was still alive and, while he apparently wasn't able to talk, he was swivelling his eyes to follow the events around him. I leaned over and said: 'I'm sorry, Ahmad, I'm really sorry.' I then started patting down his combat fatigues for any spare magazines he might have left. The image of that poor guy, lying dying on a dusty roadway, will never leave me. I was looking directly into his eyes as I desperately searched for spare ammunition clips, and it was heartbreaking. Our situation was so desperate that we were resorting to scavenging spare rifle clips from the dead or dying.

By now, there were just four of us left. All the civilian drivers were dead. It was just Yves, myself and two Iraqi security contractors, Heider and Ali. We were still fighting back but were being overrun. I glanced around the front of the truck cab to see whether there were any insurgent movements to our left when Ali, who thought he had spotted something, opened up with a volley from his AK. But the muzzle of his rifle was just inches from my left ear and the blast wave left me reeling. I initially thought I'd been shot but then realised what had really happened as my ears throbbed and my head spun. It felt like I'd been hit with a mallet and for a few seconds I struggled to get my bearings.

I turned to Ali and said: 'What the fuck are you doing?'

My head had hardly cleared when, as I glanced over my shoulder, I saw an insurgent running along the far side of our wrecked trucks. He was less than twelve feet from our position, between the side of our trucks and the main ambush position. In horror, I realised that they were now trying to infiltrate the convoy itself. If they succeeded, we were all dead men. I threw myself to the roadway and opened up with the AK, firing parallel to the ground. I don't know whether my volley hit him or not but, as I took further aim, I realised that the roadway underneath the car and trucks was soaked in a pool of leaking fuel. The tanks of all the vehicles had been shredded with gunfire and we were now sitting on the biggest Molatov Cocktail I'd ever seen. One spark and we'd disappear in a fireball.

I shouted to Yves about withdrawing to the reeds by our only open flank. We had hardly any ammunition left and our position was now totally compromised by the leaking fuel tanks and the insurgents sweeping around the trucks. But before I could even finish my warning to Yves about the leaking fuel, he was killed. He'd dropped the radio and was fighting back with his AK when, from around the rear of one of our convoy artics, an insurgent fired a blast directly at us. The shots caught Yves on his side, which was unprotected by his body armour. The same shots that killed Yves then killed Heider, who was standing by his side. The two men collapsed and fell, as they had fought, still side-by-side.

I knew I was now in the last-chance saloon. The insurgents were amongst our vehicles and we were about to be overrun. In shock I realised that there was only Ali and myself left from the sixteen of us who had started out in the fire-fight. I looked up to explain to Ali about my plan to make a run for the reeds when he too was killed. A single 7.62mm round took him in the forehead and tore away the top portion of his skull. He dropped on to the roadway like stone. In horror, I realised that I was now on my own.

There's a lot of shit written about the 'last man standing'. The truth is that there is nothing like the isolation or the loneliness of realising that your comrades are dead and you're the only one left alive on the battlefield. It's even worse when you realise you're surrounded by people who want you dead. Battles often happen in a matter of seconds and, for the most part, a soldier fights for the friends at his left and right shoulders. It's not a question of debate – you fight to protect yourself and your friends.

I'd never known loneliness like it. It was like a hammer-blow between the eyes. I stared around me and all I saw were the corpses of my friends. Guys that I'd drunk with, fought with and worked alongside for months. Now, I was on my own. And, to be honest, I figured I was going to die. The strange thing is that I didn't feel afraid. Not the slightest trace of fear. I felt angry – and had this huge sense of loneliness. I knew I couldn't afford to let them take me alive, but I was shocked by just how isolated I felt. Even when there was just myself and Ali, at least we supported each other despite the desperate plight we faced. Now there was only me and a near-empty Kalashnikov to keep me company. We had held a near-impossible position for more than an hour and a half and the savagery of the assault was borne out by the fact that someone was killed every three minutes during the engagement.

I checked the magazine in my AK and saw that I had only three rounds left. Just then two insurgents stepped out from behind one of the artic trailers and, without thinking, I fired my last shots at them. They were less than twenty feet from me and I know I hit one, but the other was probably unscathed after diving for cover. But I was now effectively unarmed and needed to get either another gun or ammunition fast. I crawled over a pile of bodies and under the wrecked civilian truck towards the roadside by the field of reeds, with my empty AK in my hands. I had bullet wounds to my face and elbow and I knew from the bruising to my chest and back that my body armour had protected me from other impacts. I spotted an old tree and, directly beside the tree, two legs from a body that was partly hanging over the roadside parapet. Miraculously, slung across the back of the dead man was an AK-47.

I slid down beside the body and tore the AK from its slung position on his back. Just as I got it free, two more insurgents ran around from the front of the wrecked truck that had been our lead artic. I raised the AK to engage them but, even as the gun came up, my brain was registering that there was something wrong. I felt my heart skip a beat as I realised that the new AK had no magazine. Without thinking, I dropped it and raised my own empty AK. Almost unbelievably, the two insurgents saw me engaging them and immediately scattered back behind the truck.

Time seemed to stand still. I knew at that very instant that my options were either to run for Sean's bullet-riddled Opel and be killed or stay where I was and be captured alive. In the milliseconds available to me, I weighed my chances and knew that, no matter what option I chose, I was going to be killed. The mental conversation with myself left little hope of escape, but, strangely, I still didn't feel any fear. If anything, I was resigned to what was about to happen. After ninety minutes of ferocious fighting and watching my friends die, I knew it was my time. But, I thought, it was better to die on my terms than to wait for what would happen if I was captured alive. I didn't even bother to pull the trigger on the empty magazine, I dropped the gun and ran towards Sean's car. I considered the run towards the car as little more than attempted suicide. Better to die like a soldier on my terms than be captured alive and put in an orange jumpsuit.

I expected to be killed by a bullet from behind as the insurgents broke cover. The burning convoy was now overrun with insurgents. I still wore my body armour and I ran as fast as my battered legs would carry me. As I cleared the truck, two insurgents crept around the front of the cab and I was so close to them I could have punched one in the face. But I startled them and they obviously didn't react for a few crucial seconds as I raced by. By now, the insurgents were swarming all over the place and that meant that their machine-gun couldn't be used for fear of killing their own men. That gave me the only chance I needed. Otherwise I wouldn't have made it more than three paces beyond the cover of the artics before being cut down.

I was about five metres from the car when I suddenly felt as if I'd been kicked in the leg by a horse. I was astonished that I had made it this far. Before I could react, my leg had collapsed under me and I crashed down on the road. I'd taken a bullet through the upper thigh, but I knew I had to make it to the car or I was a dead man. I lay on the roadway, my hand on the gaping hole torn by the bullet in my thigh, my heading resting on the stone road surface. I could see the insurgents clearing the convoy and moving towards me. Part of me was simply exhausted and wanted to just lie there. But another part of my brain was screaming to get up, to keep moving and to make it to the car.

I desperately dragged myself to the rear of the car and, just as I made it to the door, the roadway and car erupted in a spatter of incoming rifle rounds. The whole air around the car seemed to 'zip' with passing rounds. But grabbing the PK heavy machine-gun in Sean's car, I didn't even wait to aim it properly before pulling the trigger – only to hear nothing. I stared at the gun dumbfounded and realised that the ammunition belt had twisted and jammed the mechanism. I ripped open the breach and tried desperately to free the belt before it was too late.

But I knew I had only seconds left, and the belt simply wouldn't come free. I felt like screaming at the heavens in frustration – to get this far and then find the machine-gun jammed! But instinct now took over. I dived clear of the car and threw myself off the roadway down the berm towards the field of reeds. I vaguely remembered the image of the car being shredded by AK rounds as I fell off the roadway. If I'd freed the PK ammunition belt I would have hammered the insurgents – but I also know that I would have stayed in the car and that's where I would have died.

I hit the drainage ditch below the road escarpment like a brick. Every part of my body was now screaming in pain. But I had to get clear and I only had seconds left before the insurgents would be peering out over the roadway. The question now was whether I should head into the field of reeds or towards a drainage tunnel that I noticed the culvert fed into. Without hesitation, I crawled and dragged myself towards the tunnel. My leg was wrecked and I simply didn't have the strength left to make it into the reeds. And heading towards the reeds also meant that I would have given the insurgents at least fourteen feet of a clear shot as they stood on the elevated roadway.

I tore off my body armour and threw it towards the field of reeds as if to indicate that I had headed that way. I then made it to the drainage tunnel through a mire of mud, slime and fetid water. God, but the tunnel stank. I pushed and hauled myself into the darkness of the tunnel and waited for my fate. The pain then began slowly to wash over me like a giant wave. My face was numb, my head was still ringing from Ali's AK blast, my elbow was throbbing and there was a searing pain in my upper thigh. My arms were raw from glass shards and I knew I was bleeding profusely. I could see the dark waters of the drainage tunnel being stained red with little circlets of my blood.

I fought against the pain and shock and wedged myself

deeper into the tunnel. There was less than six inches between the surface-water level and the tunnel top and I forced myself to stay rigid and steady as I downed lungfuls of stinking air. Overhead, I could hear the insurgents walking up and down the roadway. They were jabbering away to each other, but I couldn't hear any signs of a pursuit for me. They sounded exultant. By this stage, I reckon, I must have been going into shock because the greatest concern I had was that, if they discovered me from the reed side of the tunnel, I would be shot through the legs and genitals. Better, I thought, to be shot from the other side of the tunnel where at least I'd die from a head wound.

It never even entered my head at this point that I could still be captured alive. I was completely unarmed. I didn't even have a bayonet or pen-knife with which to defend myself. Silently, I cursed myself for not even wearing a small knife tucked into my boots. Fuck, you're around long enough to know something as basic as that, I silently seethed. Now, if I was found, I had only my bare hands to try and fight off the insurgents. I knew a knife wouldn't be much use against an AK, but at least it would give me the option of fighting hard to make sure I wasn't captured alive and forced into an orange jumpsuit. But it was too late to worry about a knife now. I knew that the tunnel was more than likely going to be my grave.

I began slipping into and out of consciousness. But I remember hearing a car race up to the ambush point, wait about five minutes, and then drive away again at speed. The next thing I remember is the sound of explosions, which rocked the tunnel in which I lay. I guessed that the insurgents were throwing satchel-charges into the trucks to destroy them and their cargo, which the ambushers had by now realised was absolutely worthless to them. Then came the worst sound of all – the firing of volleys of two shots, each in quick succession. With two execution shots to the head, the insurgents were making sure all the convoy personnel – from civilian drivers to security contractors – were dead.

But I still couldn't hear any sound of a search or hunt for me. As I lay in the stinking water, it dawned on me that there were so many bodies strewn all over the roadway the insurgents had probably lost track of me. They may even have thought that one of the bodies in Sean's car was mine. Even a blind man could have followed the trail I had left behind me, dragging myself through the culvert to the drainage tunnel. And I knew that my body armour would only throw them off the scent for so long. But I began to suspect that the insurgents were probably also a little shocked at how effective their ambush had been. In their wildest dreams, I don't think they could have imagined themselves overrunning an entire convoy. But that was precisely what they had done.

Slowly, the sounds began to fade and I realised that I had got away with it. My only problem now was to wait until darkness and figure out a way of making it back to Coalition lines. That would be no easy task given the physical state I was in. And I knew that any locals here would either kill me themselves or gladly hand me over to the orange-jumpsuit gangs. I needed a weapon – but how was I going to get one now? I vaguely noticed the sound of a helicopter flying overhead. Then the sound of the rotors faded and, a short time later, I heard the distinctive roar of approaching jet engines. Within seconds, the whole earth seemed to shake and I thought the tunnel would collapse around me. The world seemed to dissolve in a wave of dust, smoke and splashes of slime.

US jets had dropped bombs to clear the area of insurgents and the shock wave swept over the area like a tsunami. Then, as the dust began to clear, I began to realise that the water level in the tunnel was slowly dropping. Whereas I once had just six inches of air in which to breathe, I found the tunnel was now almost dry. I presumed that the US jets had wrecked the local drainage system with their bombs and the drainage water had now been diverted away from the tunnel. I gazed around me and realised that there was light coming from only one side of the tunnel, the side by the field of reeds. The other end of the tunnel was in darkness because it was backed by a high ditch.

I started to pray like I had never prayed before. Before I'd returned to Baghdad just a few weeks earlier, I had asked my mother for Rosary beads. I wore them around my neck but they had somehow been broken during the early stages of the ambush. But I prayed anyway and, in that stinking tunnel, I prayed without hope or expectation. Suddenly, I felt that I wasn't alone and that something was guiding me out of the tunnel. I knew, deep inside, that I had to get out of the tunnel right away so I began to edge my way towards the exit.

I knew I had to get out of there before I was found by locals investigating the scale of the bombardment damage. But it was a slow and painful exercise. My whole body seemed to protest and it took me several minutes to finally push myself to the tunnel exit. As I poked my head out into the fading sunlight, I was shocked to look up and find myself staring at the barrel of the 25mm chain-gun of an M2 Bradley armoured personnel carrier. The gun was highlighted against the darkening sky and was trained out over the elevated roadway towards some unknown threat. I heard voices overhead on the roadway and realised with a mixture of relief and joy that they were American.

With a start, I knew I had to make a move if I was to be rescued. I started shouting, 'Friendly force, friendly force' – but it was more like a hoarse croak than a shout. There was no way it could be heard over the rumble of the Bradley's engine. I crawled and scrambled up the berm on my hands and knees before finally staggering out on to the roadway with my hands held over my head. I kept calling, 'Friendly force, friendly force', as if it was a mantra that my very life depended upon. I must have looked like something from a B-budget horror movie. My face was caked in blood, slime and dirt, my fatigues were soaked in blood, my arms were black from the tunnel and I was trying to balance myself on my one good leg, as my other leg screamed in protest if I put any weight on it.

The instant I got out on to the elevated roadway, I saw the 25mm gun traverse slowly in its turret and point directly at my chest. One blast from that chain-gun was enough to cut a car in half – it would reduce a human being to mincemeat in seconds. With a start, I realised that the M2 was the second last vehicle in the US rescue-evac mission. If I'd waited just two minutes more in the tunnel, I would have missed the entire convoy. That guidance in the tunnel to get out and get on to the roadway had saved my life. Suddenly, it was all too much. Fuck this, I just can't take any more – enough is enough, I thought to myself as I unceremoniously collapsed on the roadway in a heap. I hardly even noticed the pain of the impact, I was so exhausted. Seconds later, I felt the barrel of an M16 rifle nudge against the back of my head. The US troops weren't about to take any chances given the charnel house they were viewing. 'Friendly force, friendly force,' I kept whispering. I heard a hurried conversation over my head and the bark of an order from an officer inside the Bradley. Then, slowly and carefully, two young US Army privates rolled me over and gently checked me for weapons. 'You're okay, buddy, you're safe now,' one young American GI smiled at me. But I saw the other trooper grimace as he stared at the condition I was in and at the smoking ruins of what had once been my convoy.

I forced myself to crane my neck sideways so I could look back at the convoy. The bodies of our Iraqi drivers and contractors lay sprawled around the still-smouldering hulks of our trucks. Every single vehicle had been torched and evidence of the fierce fighting was everywhere. The entire roadway was littered with AK shells and all the vehicles looked like giant sieves, they had so many bullet holes. My Toyota was virtually unrecognisable – if it hadn't been for the position of the car, I wouldn't have been able to identify it. The US patrol was cautiously examining the ruins and exercising maximum care as it wasn't uncommon for insurgents to leave booby-traps and IEDs behind at ambush sites. But in this case they must simply have run out of time and had to get clear of the area before the US patrol arrived.

I was almost too exhausted to register what I had just survived. I reckon I was also in a mild state of shock, and the only thoughts that seemed to occupy me was where Sean's and Yves's bodies had gone and what had happened to Denis. I quickly realised that there was no trace of Sean or Yves – and suddenly that car which I had heard speeding up to the ambush site while I was edged in the pipe sprang to mind. Had it been used to remove their bodies? I simply didn't know. All my over-stressed brain could grasp was that the convoy I had escorted out of Baghdad just a few hours ago was now a blackened, scorched mangle of men and machinery.

The strangest thing was that the hamlet which just over an hour ago had been like a scene from Dante's inferno was now quiet and peaceful. In fact, there was absolutely no sound whatsoever. All I could hear was the rumble of the Bradley's engines, the crackle of the slowly dying fires by the trucks and the dull, ringing sound in my own head. I couldn't see a single living soul around the buildings, and there didn't even seem to be a bird in the sky. It was almost as if the entire area had been cursed and abandoned.

I reckon that the US troopers were by this stage very worried about me because one of them kept chatting to me, almost as if to try and keep my spirits up. But I was almost beyond emotion, as if my whole system was threatening to shut down from a total overload of stress, pain and adrenalin-charged feelings. I was alive, but my friends were dead. I had survived – but at what cost?

As I was helped towards the waiting Bradley, I realised that, for the second time, I'd been spared the rule of three.

13

'BUT YOU'RE DEAD!'

I half-walked and half-staggered towards the second last Bradley armoured personnel carrier. The two US troopers who had been assisting me had obviously come from the last Bradley in the convoy as they immediately ran back towards it, leaving me by the second-last vehicle. I leaned, totally exhausted, against the armour-plated side of the Bradley and banged on it to be allowed in.

But I didn't realise how the Bradleys operated and I was standing too close to the rear loading ramp for it to be safely lowered. By this stage, I was beyond all logical thought and all I could see was that I was the only person still left exposed on the roadway. I banged repeatedly on the side of the Bradley, shouting, with increasing emotion: 'Let me in, for fuck's sake, open the door and let me in.' This farce must have continued for several minutes before the trooper who had helped me on the road jumped out of the last Bradley, ran forward and dragged me aside so that the armoured ramp could be lowered to allow me in. I was instantly dragged in by another US trooper and, for the first time in hours, I was safe. But I felt nothing. I had expected to die on that roadway and now, suddenly, I was alive. I just didn't know what to feel or how to react.

Those US soldiers couldn't have done more for me. They treated me like a wounded comrade. A female medic inside the armoured interior of the Bradley gently began putting field dressings on all my wounds. The medic had to cut the filthy clothes from me, and I can assure you I was far beyond the point of being embarrassed. She was very concerned about the fact that I had been in the fetid water for so long and she immediately started me on an intravenous drip, laced with a cocktail of antibiotics.

I have no idea how much time passed before the patrol finally began to move off. Its only route back to base led right through the ambush site. The commander of the patrol conferred with his troops via the intercom and I could hear the orders being given through the headsets near me. But I was still shocked when the commander asked whether there was room for bodies in the rear of the Bradleys.

Instantly, the young trooper who had dragged me into the Bradley looked at me and I shook my head as if to say: I don't want to see any more bodies today, thanks. He conferred with his commander in the lead vehicle but I knew from his posture and the look on his face that he had no intention of leaving the safety of the Bradley's interior. I had seen the full horrors of what had happened out on that roadway and I didn't blame him for wanting to stay inside the APC.

By this time I was totally wasted. I had lost a lot of blood, I was aching from multiple wounds, I was exhausted from the adrenalin-surge of the combat and, worst of all, the medic had put me on morphine which was now beginning to kick-in. I eased my battered body up on one elbow and looked over at the trooper scanning the roadside through the gun-ports. 'How many more did you get out?' I asked. He tore his eyes away from the carnage outside and tried to avoid looking me in the eye: 'Buddy, as far as I know, you're the only one we got out, I'm sorry.'

I sat back and the shock now fully hit me. My first reaction was to wish that I had died too. I didn't want to be the only one of the team to make it out – in my drug-clouded mind it just didn't make any sense. It would have been better to go down with guns blazing, I thought. Slowly, the combination of exhaustion, blood-loss and morphine got the better of me. My head was wrecked. I didn't know what to think or what to say. All I could do was lie there in a heap. The only thought that seemed to crystallise in my mind was that maybe it would have been better to die than to survive and have to live with the memory of what I had seen on that roadway.

But the young American troopers kept talking to me, assuring me that everything would be okay and that I was now safe. When they realised that I was Irish, the US soldiers immediately began asking me about where I was from and how I had ended up in Iraq. They also told me that, as I had guessed, the air-strikes had been launched because a reconnaissance chopper had been shot at by insurgents from the ground. I learned that the US patrol was responding to a request for assistance from both their

observation post in the area and from our security compound back in Baghdad. Yves's desperate plea for relief had
got through, but the cavalry had arrived just too late. One
US trooper told me that the entire area was now rife with
insurgent activity and all their patrols had been assigned
armour support. Our old Opels and Toyotas had never
stood a chance.

The US medic seemed amazed that I wasn't more seriously injured, but she was concerned about the wounds to
my face and leg. Initially I must have looked like I had been
dipped in blood, and my face was grimy from the combined
effects of heavy gunfire and explosions. I was a mess and I
knew I must be stinking to high heaven. But I was alive.

By now, the US patrol had cleared the ambush area and
made it back to one of their forward control bases where a
full medical team was on standby. I was gently lifted out of
the Bradley, placed on a stretcher and loaded on to a four-
tonne open-backed truck with armour-plating around its
sides. The truck then bumped and bounced its way across
the roads to a major US base where there was a field surgery
team. I was taken into the casualty ward and the doctors
began a full assessment of my injuries. I have only a hazy
recollection of that time because the morphine was now well
and truly taking hold. But I do remember a US Army Chaplain coming into the surgery and asking me was I Catholic. I
replied that I was and that I was from Ireland. This seemed
to fascinate him and, after a few minutes of stumbled conversation from me, the Chaplain turned to me and asked
whether I was ready to receive my Rites?

The drug-induced fog immediately parted and I sat bolt upright on the surgery table and asked: 'Jesus, do you mean my Last Rites?' The Chaplain, who was obviously shocked by my reaction, immediately tried to calm me and said: 'No, no, no, son. Your Rites for the Wounded. They're special prayers for casualties. They're not your Last Rites. You're going to be okay.' He was a decent guy and stayed with me until the doctors were ready to begin their work.

Sometimes being Irish has its advantages and the doctors seemed fascinated by where I was from and what I was doing in Iraq. They also seemed determined to get me to discuss the phenomenal popularity of Irish stouts like Guinness and Beamish – which, even at the time, seemed a strange conversation for someone who was lying wounded and drugged on an operating table. Maybe they were just trying to keep me conscious.

I never lost consciousness but the morphine was making my mind wander from the horrors of the ambush to something as innocuous as going for a pint with my mates in Cobh. Eventually, after the doctors had completed their preliminary treatment and a full assessment of my condition, I was told I was listed for medical evacuation. This meant being transferred to Landing Zone Washington, a major military transport hub for flights into the maximum-security Green Zone in Baghdad. I was to be transferred by US Marine Corps ambulance, in this case an armoured Humvee.

By now, news of what had happened was reverberating around Iraq. Convoys had been hit before and had suffered serious losses. But we had been virtually overrun and annihilated. Almost an entire PMC security team had been wiped out. We ranked as amongst the most experienced personnel 'in country'. If the insurgents could take on and kill a group like ours, then what did the future hold for other convoy missions?

At that point, those concerns were way beyond me. The pain was now beginning to hammer me with a vengeance. My elbow was throbbing, my face felt as if it was on fire, my leg was numb, my arms were stiff and I was wincing from a pain in my back. My neck and face were sore and stiff, and later I learned that they had been peppered with shrapnel fragments. Two years later, I was still getting shrapnel taken out of my neck. Back in the Bradley, when I told the female medic about my back, she examined me and let out a long, low whistle. When I asked what was wrong she told me I had a deep bruise, the size of a saucer, right along my spine. It was left by the impact of a bullet being stopped by my body armour.

I managed, as best I could, to thank the US personnel for saving me and they all wished me well. I owe them my life and it's a debt I will never forget. When I arrived at LZ Washington for my air transfer to hospital in Baghdad, the drugs were beginning to wear off and my oldest addiction kicked in with a passion: I was absolutely burning for a cigarette. As I was unloaded from the Humvee by two huge Marines, I turned to one and asked was there any chance of a cigarette? Without saying a word, this monster of a Marine waited until my stretcher was put on the ground and then leaned over and placed a fag in my mouth and gently lit it for me. I was just about to whisper my thanks when, with horror, I realised that the cigarette was menthol. 'What the fuck! Look at the size of you and you're smoking menthol cigarettes, you wanker?' I ranted. 'Why aren't you smoking Marlboro or Lucky Strike?' The giant Marine's face split into a grin and he laughed as he walked away.

I was then put into a medevac helicopter, a giant Sikorsky Chinook. I was one of the first casualties loaded and I was assigned to a stretcher 'hook' just inside the tailgate of the huge twin-rotor chopper. When it was fully loaded, the pilot then left the ground in a combat take-off – a near vertical ascent aimed at foiling potential snipers or missile launches. But for me, lodged just inside the open rear ramp of the chopper with my head just inches from open sky, it was a terrifying experience. One minute my head was on the Chinook's floor just two or three feet from the ground, the next minute I was staring into open sky with the ground several thousand feet below. It was like undergoing a parachute jump in reverse. Shit, I didn't survive that ambush to die by falling out of a helicopter on my way to hospital, I thought – but it was a great adrenalin rush all the same. And after a short ride, the Chinook flared and landed at a military hospital within the heavily-fortified Green Zone in Baghdad.

In Baghdad I was in control of my senses but I was also shocked by what I'd just gone through. I'd driven by this hospital so many times on protection duties and now here I was a patient, swathed in bandages almost from head to foot. The doctors gave me another full assessment and ordered a whole battery of tests. At one point, I remember looking around and seeing doctors and nurses swarming all over me and I thought it was like a scene from *ER*. But I knew I had been very lucky to escape with such light injuries, and even the doctors were amazed. I had a gunshot wound to my left cheek, a bullet had gone straight through my right thigh, another bullet had torn through my right elbow, my left eardrum was completely perforated, I had shrapnel wounds to my neck and back, lumps of flesh had been scraped from my face, hands, arms and chest.

I had thought the cheek wound was the most serious but one doctor insisted that my leg wound had been inflicted by 'a magic bullet.' As I stared at him in bewilderment he explained what he meant. 'The trajectory of that bullet meant that it should have travelled up through your thigh, into your groin and out through your stomach,' he said. 'I don't know what made that bullet rotate the way it did, but it shouldn't have exited your leg where it did. And it missed your femoral artery by less than an inch.' A wound to the groin or femoral artery is something most infantrymen dread because if the artery is nicked you can bleed to death in a matter of minutes.

I knew I couldn't take this for long and I was getting increasingly pissed off, probably due to the fact of shock being replaced by anger and my adrenalin levels subsiding. I had escaped the worst ambush in Iraq to date and here I was sitting in a hospital with only flesh wounds. About six hours later, I called one of the doctors and asked whether I could go back to my base and recover from my wounds there. I suddenly had a desperate need to get out of that hospital as fast as possible. I think he understood what I was going through because he re-examined my wounds and said they were superficial and I could technically sign myself out of the hospital.

He quickly organised a pack for me containing medicines, pain-killers, bandages and special cleaning agents for the wounds. But he stressed that for both my leg and hearing I might require further specialist treatment. Of course I'd forgotten that I was as naked as the day I entered the world because my filthy combat clothing had been shredded by the US medic. So, the doctor organised a pair of flip-flops, a pair of shorts and a T-shirt for me. They were massively over-sized, but they were sufficient to get me out of the hospital and that was all that mattered for me.

I checked myself out and, because we were in the Green Zone, I knew precisely where the base operations control centre offices were and that there would more than likely be a Hart Group agent there. So, slowly and painfully, I hobbled towards the security compound in my shorts, flip-flops and tent-sized T-shirt. It may sound crazy, but for my sanity I needed familiar surroundings.

When I limped into the compound, there was absolute silence from the other contractors. They all went pale. One looked at me in shock and gasped: 'But you're dead.' Only now did I learn that, after the ambush, I was posted as missing, presumed dead. We all carried emergency transponders and, when triggered, these showed up on a computer screen in the firm's head office. But I also heard that there had been a problem with all the transponders the morning before the ambush and, given the circumstances, they all presumed that I'd died along with the others. A senior manager eventually took me to temporary accommodation and it was the first and only time in my life that I slept with the lights on in my bedroom.

The next day, I was taken by an unmarked car sent by Hart headquarters back to my staging compound – and, placed beside me on the rear seat, was an AK-47. I looked at the gun, looked around me and thought: Fuck, not this again! But this was Baghdad and once we were out of the relative safety of the Green Zone, we had to carry arms, even the wounded.

I met Denis later in the Hart HQ. He had made it to safety at the US Army compound our convoy had passed. He stormed up to me and said: 'Where the hell were you? Why the fuck didn't you come back to help us.' He obviously thought that the front part of the convoy had got clear and that only his part had been trapped. I didn't take the question well and roared back: 'Where the hell was *I*? Where the hell were *you*? I'm the only one left.' The stunned look on his face made me realise that, until now, he hadn't realised what had happened to us. When it suddenly sank in that I was the only survivor of the entire convoy vanguard and that both Yves and Sean were dead, he sat down on a couch and reached for his cigarettes. Like myself, all he could do was smoke. Neither of us felt much like talking for a while.

Eventually, we both cooled down and exchanged stories. The worst aspect for him had been the fact that his family had been informed he was missing. Officials had also attempted, mercifully without success, to pass the same message to my folks. Incredibly, the telephone number for my family had been mistakenly transcribed so they couldn't contact them despite repeated attempts. The first my family heard about the lethal ambush was when I was able to ring them and assure them that everything was okay. As far as my parents were concerned, it was like our regular Sunday chat. But it wasn't so easy for Denis whose wife back in France was understandably distraught. She had been told that her husband was missing and presumed dead – and she now refused to believe that Denis was, in fact, alive and well and trying to speak to her down a crackly telephone line from Baghdad. After several minutes of trying in vain to persuade his wife that he was okay, Denis handed me the telephone in exasperation and told me to confirm his identity. I took the handset and chatted calmly with her, assuring her that Denis was safe and well, but I could tell from the tone of her voice that she was emotionally shattered. Eventually, her emotion started to get to me and I threw the phone back to Denis and said: 'Tell her yourself.'

Once we were comfortable in the security compound, a tasking official from the Hart Group arrived to get an outline briefing on what had happened. I explained that I'd seen Yves, Sean, Arkan and Wisam die. I also tried to give them as many details as I could about what I'd seen happen to our other Iraqi contractors. I told them, as best I could at that stage, the precise sequence of events that led to the slaughter because I knew it was very important that Hart understood exactly what had happened so that other teams could avoid our fate. I hoped there were lessons to be learned, but, given the appalling security situation in Iraq, I had my doubts.

Inch by inch, we went back over the route the convoy had taken, how the truck had broken down, the contact with the US patrol and finally how I had spotted the ambush. Hart were fantastic – they were very understanding and they fully took on board what we had just gone through. I learned, for the first time, how Denis had made it to the US outpost and pleaded for help for us. It had taken him just seven minutes to get from the ambush site to the outpost and the troopers on duty could see and hear the firefight that was wiping out our convoy. But they simply told Denis they had no reinforcements available as their patrol was already out on a mission.

Denis and his Iraqi colleagues, despite having military identity tags, had ended up being taken by the US military to a prison complex at Ramadi where they were interrogated in detail about what had just happened. When the US officers were finally satisfied as to their story, Denis and his team were fed and given accommodation until Hart officials could collect them. It was only as Denis and I told our stories to the Hart officials that we all began to realise that the heavy traffic out of al-Habaniyah that day as our truck had broken down was probably caused by insurgents clearing the village in preparation for the ambush.

The next day, Denis and I had another debriefing and this time the Hart official had brought copies of all the photo-identity passes of our Iraqi contractors. I had to take the

copies and go through each one, confirming that I had seen
one die, or that I had no information on what had happened
to another. Only then did it strike me that I had actually seen
most of the group die right before my eyes. As if I hadn't
already gone through enough of an emotional roller-coaster,
I would now have to assist in dealing with the relatives of
the Iraqi security personnel who had been killed. The next
day, a group of about thirty relatives arrived at our security
compound – fathers, brothers, cousins, sons and uncles.
They all wanted official news on what had happened to their
loved ones. Another printed page had been organised with
all the photo identities of the Iraqi security contractors and
beside each we had to indicate 'Killed in Action' or 'Missing
in Action'.

I felt honour-bound to try and help. These were the relatives
of the guys I had fought alongside for my life, and they had
fought every bit as bravely as Yves, Denis, Sean or myself. The
tragedy was that they just didn't enjoy the luck that Denis and I
had. I also knew that, if they were my relatives I would want
them to hear the news from a comrade and not from a 'suit'. It
was my responsibility. I was there and I saw them falling, one
by one. But it was heartbreaking to see the pain and anguish
after I confirmed, in most cases, that their son or brother was
dead; the relatives would start wailing and mourning in
Arabic. Not surprisingly, the emotion of the procedure finally
got to me – and even now it's something that I struggle with.
The years with the Legion and with Hart in Iraq had taught me
how to keep my emotions under strict control, but the sheer
horror of this situation was something beyond the ordinary.

But I still insisted, through my interpreter, that each and every family be told that I considered it an honour to have served alongside their loved one. It was scant consolation for what they were going through. Most hadn't even got the bodies of their loved ones back.

Afterwards, all I could do was go and find a quiet spot in an empty room, sit down on the floor and chain-smoke. I had reached the point of emotional overload and I knew, deep down, that I couldn't take much more. I had done my duty. I knew I couldn't have done anything more to help my comrades and I knew that my survival was entirely against the odds. But I was alive. And that was a difficult concept to grasp when you weighed it against the loss of so many friends and colleagues. What made it all so much worse was understanding, even in a small way, the terrible grief that these families were now going through. I felt a little like Jekyll and Hyde – part of me was so grateful just to be alive but another part wished that I had fallen alongside my friends. It was total mental turmoil. And I kept asking the question mouthed by soldiers for as long as man has carried swords, spears and shields: Why me?

The Hart headquarters is part of its own little enclave in Baghdad and the streets are shut to all except security personnel. Everything is very carefully guarded and, once inside the compound, you can relax a little. But then I was told I had to go back to Abu Ghraib until our security project there was shut down. Leaving the Hart compound and going back to Abu Ghraib felt like being forced back into the mincing machine. But I had to go, and in a way it was a relief because my security responsibilities at least helped keep my mind off what had happened outside al-Habaniyah.

Denis and I were greeted at Abu Ghraib like returning family. I felt a huge emotional boost from seeing the likes of Geoff, Pete, Ritchie and the rest of my contractor colleagues. Both Denis and I were deeply moved by how happy the lads were to see us safe and well. From the entire convoy, there were only four survivors: Denis, myself and two Iraqi contractors who had been at the very rear of the convoy. And while the other contractors were clearly delighted to see Denis and myself, I knew that the death of Yves and Sean hung heavy over everyone. After a few hours, I realised that most contractors regarded Yves with something akin to awe and now they worried that if someone with his skills and ability could be killed, what chance had they? We had all gone to work at Abu Ghraib knowing that it was the worst possible security detail in Iraq. And now that detail had got even worse. No-one was sorry to leave Abu Ghraib and, within twenty-four hours of my arrival back, the entire security team had moved in phased convoys back to Hart headquarters in the Green Zone.

I don't remember much about that night because I got drunk with a vengeance – as much to try and forget the horrors of al-Habaniyah as to share the happiness of my friends at seeing both Denis and me alive. Once again, Hart were fantastic. They knew the pressures we'd been under, they knew we needed to let off steam and they did their best to facilitate that. Over the next few days, the entire Abu Ghraib team was rotated out and I had to endure a car journey to

Basra. On a good day this is a nightmare journey because the route is littered with ideal ambush spots. In my condition, I thought this road to hell would never end. And, as if I wasn't suffering enough, Mother Nature decided to brew up a sandstorm to make the journey even longer. To complete the agony, one of our cars broke down about 30km out of Basra and we had to wait for a relief team to come out and meet us. The only good part of the journey was that I got to travel with Stevie K., an Irish lad with a fantastic sense of humour who was a veteran of the British Army, a great guy and incredible company.

The welcome in Basra more than made up for the terrible journey. The security situation was nowhere near as bad in the south as it was in Baghdad and all my old friends here had heard about what had happened at al-Habaniyah. That night, they threw a party in my honour with enough booze to float a battleship. It meant the world to me that these guys, some of the toughest soldiers I've even seen, would try and show how pleased they were that I'd made it out alive.

The next day, a South African guy called Darrell, a trained medic, changed my dressings and checked my medication. He had also made an appointment for me at the British military hospital in Basra to see if I was fit enough to take a civilian flight home rather than a military transport flight to Germany as one of the so-called 'walking wounded'. The doctors at Baghdad hospital told me that I needed follow-up specialist treatment and that this would have to be carried out back in Europe. I was given the choice of being treated at a private facility in Germany or returning back home to

Ireland. Hart would look after the costs of the treatment irrespective of where I opted to go. I was also advised that word of the ambush had reached the media and, because of the heavy loss of life involved, it had made major headlines throughout the world. Even the fact that I was Irish had also emerged, and it meant that my name eventually reached the newspapers back home. I knew this wouldn't be easy for my family, so one of the first things I insisted on was getting in contact with them again to reassure them I was okay.

Luckily, the British military hospital cleared me for a civilian flight home so I travelled with the rest of the Abu Ghraib security team by road to the Kuwaiti border. We crossed without incident and were taken straight to the local Hart security compound where our flights would be arranged. Two days later, everything was organised and we were taken to Kuwait City for the direct flight back to Heathrow. Hart had even gone so far as to book me a first-class ticket, though the rest of the security team were back in economy. I loved the luxury and the miniature whiskeys, but I would rather had been with my mates. I also knew that I was out of place – the businessmen around me weren't my kind. Mind you, I must have scared the shit out of them being swathed in bandages and with half my face swollen.

I had told Hart I wanted to get my medical attention back in Ireland which would allow me to see my family and rest up for a while. I knew I needed to get out of Iraq and just take time to sort things out. I was a mental wreck at this point and I knew I desperately needed space and time to take stock and get some perspective on what had happened to me. But coming home to Cork wasn't quite so easy. For a start, the sheer scale of the ambush and the loss of life involved had made front-page news back home and my parents had been taking phone calls from reporters for almost a week before I arrived back.

I flew from London to Cork and my parents were waiting on the tarmac to greet me. Because my dad is a former policeman, he had the contacts to make our transit through the airport as fast and painless as possible. It was fortunate he did, because the airport terminal was swarming with reporters and photographers all determined to get a 'scoop' for the next edition. My mother was appalled at the state I was in, but she was thrilled that I was home and safe. I think it was only now dawning on them both just how fortunate I was to be alive. Thanks to my father's contacts we even had two motorcycle outriders to escort us from Cork back to Cobh.

News of my ambush escape made the *Irish Independent* and *Irish Times*. Because I was a former member of the Foreign Legion, there was even coverage of the incident in France. Throughout it all, the Hart Group tried to provide whatever support I needed, and they gave me good advice about staying out of the limelight and focusing on my recovery. It would have been easy to go on television and talk about Iraq. But the memories were too fresh – only a few days before this I had been socialising with some of these guys and now their families were having to base life-insurance claims on affidavits sworn by me that they were actually dead – their bodies hadn't been recovered and some of the families had turned to me to provide the confirmation of death.

It's not an easy thing to do for guys who were your friends, your workmates and who had spent weeks and months with you. These guys had chatted about their families, their wives, their children. Now those same wives and children were looking to me for answers. The problem was I didn't have any answers. I couldn't escape the thought that, at the very least, my friends deserved a decent burial. But they didn't even get that. I can only presume that a barren bit of Iraqi desert is the resting place for both Yves and Sean. So, understandably, the last thing I was concerned about at that stage was doing media interviews.

Reporters eventually tracked down my father and spoke to him about my return home. Ironically, I was back in unwanted headlines only a few days later when it emerged that I had arrived for treatment at Cork University Hospital (CUH) in the middle of a beds crisis in the Irish healthcare system. Once again, reporters tracked down my father and spoke to him about what was happening with the delays in my treatment. In the end, it took three days to get a bed at the hospital. While I normally shudder at the prospect of medical treatment of any kind, I was relieved to get into CUH because by now my face was swollen and numb, my leg was very painful and I couldn't use my right arm because of the elbow wound.

The doctors and nurses in CUH were great. They slowly sorted out my injuries and took steps to address the pain levels. But they warned me that a full recovery from some of the wounds would take months to complete. The wound to my face was so serious that I'd probably need plastic surgery some time in the future. My friends also rallied round to a tremendous degree – and, slowly, I began to heal both inside and outside. One incident at CUH stands out in particular: I was walking out of the hospital to snatch a quick smoke. I was still preoccupied by the ambush when I bumped into a guy I knew. He seemed very down and, when I asked what was wrong, he said his mother was dying. In that instant, I knew I had a lot less to worry about than he did and a slow flickering of gratitude for my survival began to light up in my mind.

I knew all of this was very hard on my folks, but they handled it as best they could. My father had been plagued with calls from reporters asking where I was, how I was feeling and would I do an interview? He dealt with them all very well; even so, a few papers still managed to get the story of the ambush totally mixed-up. I was beyond caring at that point. I'd heard that Yves's mother was preparing a special funeral service for her son in Paris. She never got Yves's body back and the family were forced to bury an empty coffin. That fact still troubles me. Yves was such an incredible character, such a tremendous leader – he really did deserve better. I also heard, through Hart, that Sean's family wanted to contact me. That wasn't easy either. What do you say to someone in a situation like that? How do you explain: Yeah, I saw your husband or your Daddy killed and this is exactly what happened?

I found the entire situation surreal and it was starting to

wreck my head. I think only someone who has been through the emotional highs and lows of combat can fully understand what it is like. How do people think you'll feel when they congratulate you on being alive when all you can think about is reliving the last moments of the friends who died by your side? I'm not a rude bastard, but I've cut a few people short who've said to me that, 'Tomorrow is a better day.' It's not a better day for Yves or Sean or Arkan, and the mental images of watching people die whom you've worked and socialised with don't fade that easily. I learned the hard way that death in an ambush is not a noble or a glorious thing. And the last thing I needed for dealing with the aftermath of this carnage was a cheap cliché.

In the end, I felt I had to do something about all the mistaken details that were being circulated about the ambush. To be honest, it was all coverage I could have done without. From being a low-profile operator all my life, I suddenly found being in the limelight desperately uncomfortable and slightly bewildering. I had repeatedly refused all requests for interviews or photographs. The last thing I wanted was for my photo to be splashed all over the media so that I'd be a walking target.

But the coverage of the ambush in the Irish and British media was beginning to piss me off. After it was reported that there was a delay in me receiving treatment at CUH due to the beds crisis, several people rang local radio stations complaining that I was only a mercenary and didn't even deserve a bed in an Irish hospital. One person opined that I should be immediately sent back to Baghdad for medical treatment there. And, needless to say, most papers got the details of the ambush completely wrong – and one paper went so far as to explain that I survived only because I had given my rifle to my Iraqi guard and made good my escape. Reading reports like almost hurt worse than having been shot.

So I agreed to take a telephone call from one of Ireland's top radio programmes, the Gerry Ryan Show on 2FM, just to clarify a few issues and put some of the facts on record. The interview was brief and to the point. I wanted the basic facts of the ambush aired so that people would know the real story. 'Everything that happened was geared towards getting us out safe,' I explained. 'When we actually got hit first I gave my side-arm to my driver and if I'd had that firearm at the very end I would have continued to fight and I would have been killed in that position when it was overrun.'

I briefly explained the sequence of events surrounding the ambush and just how lucky I was to be alive. 'If there had been a magazine on the weapon I took off a body when I left my first position I would have been killed fighting there,' I admitted. When Gerry asked me what my next plans were, all I could reply was: 'I will take a couple of months and see what the lie of the land is. But I just might go back [to Iraq].'

Those were tough weeks – both for me and for my parents. I have heard it said that years and maturity ultimately bring insight, and I'm beginning to understand the wisdom of that old cliché. For years I lived my life the way I wanted, and, to be honest, I didn't give a damn about the concerns or worries of others. I knew what I wanted to do and I was pretty single-minded about doing it. In fact, I regarded that commit-ment as one of my personal strengths, rather than selfishness. So I didn't give too much thought to the feelings of my family and friends or the implications of my career for others.

Now I began to understand just how tough it was for the people I love most on the 'home front'. My life has been a learning curve and I bear the scars to prove the point. There's not too much I would change about my life, but I do feel guilty about the fear and worry that my military career caused my parents, my sisters and my friends. It's a guilt I suppose I share with soldiers all over the world who are con-cerned about their loved ones back home. I've learned that fear comes in many different forms and the fear endured by families thousands of miles away from the front line is all too easily ignored.

My mother worried about her only boy being in the Middle East. 'When Pádraig said he was going to Iraq it was a massive anxiety,' she told people after the ambush. 'I think I prayed for him every single day that he was out there. I even asked the Bons Secours nuns that I work with to pray for Pádraig and I firmly believe that those prayers played a part in getting him home safe. I prayed to the Lord and I prayed to Pádraig's Guardian Angel to keep him safe. No-one can persuade me that there wasn't some kind of divine protection in Pádraig surviving the ambush the way he did and getting back home safe.'

My father said his first concern on hearing about the ambush was for the scale of my injuries. He told people: 'I heard that he had been shot in the face and leg. I also knew that he'd had his ear damaged. But we just didn't know how bad the wounds were going to be. Of course we were worried.'

I was sore and bruised but I had escaped incredibly lightly from what had been one of the worst ambushes ever mounted by insurgents. More than twenty-two people had died and most of them were my friends. I was hit by a tidal wave of relief that I was alive – and guilt because I had survived and Yves, Sean, Wisam and Arkan had not.

Hart advised me, very delicately, to think about counselling. I knew they were right, but I also knew that the people I wanted to share my feelings and emotions with were my friends. If I was going to delve deep inside myself, I wanted it to be with people I trusted and who really knew me.

I also knew that alcohol wouldn't help, but I drank anyway. At times, I think, I drank simply to forget. At other times I drank just because there didn't seem to be anything better to do. And, slowly but surely, I spoke to my friends back in Cobh about what had happened, and I began to deal with the pain. I knew there was nothing more I could have done – in hindsight, perhaps the convoy should have turned back for Baghdad after the truck broke down, but that wasn't my decision to make. I'd fought hard to protect the people in my care and I'd almost been killed making the only decision left open to me about getting our heavy machine-gun into action. I don't know why I survived. But I believe that, whatever about fate and destiny, it just wasn't my time to die.

Still, accepting that fact wasn't so easy. My friends in Iraq had briefed me on the aftermath of the ambush, and, with horror, I learned that photos of the ruined site had been placed by the insurgents on the Internet. The most stark image was of the shattered hulk of the convoy lorry, literally torn apart by the combination of bullets, explosive charges and fire. When I looked at the picture, I realised that it was nothing short of a miracle I was still alive.

I suppose it was inevitable that there would be some kind of backlash against me from the public. Most of it revolved around claims by some people that I was a mercenary, which I wasn't, I was a security contractor. They also made a lot of noise about the fact that I was Irish. A few anti-Iraq war websites ran graphic pieces on the ambush, and, reading between the lines, you almost got the feeling that they were probably sorry anyone at all had survived.

But I was alive, I was recovering and I knew that, despite the horrors of what I'd seen on the road to al-Habaniyah, I was still not prepared for a life on civvy street.

14

THE HELL OF HAITI

After al-Habaniyah I'd had enough of Iraq, though the Hart Group offered me a return contract if I wanted it. After the ambush, they were fantastic and simply couldn't do enough to try and help me. From start to finish, they were a class act to work for. But the point about working is that some day you'd eventually like to be able enjoy the money you've worked for. And a great contract in Iraq is not much good to you if you're dead.

So I opted to travel another road. One of the lower-profile elements of the security business is bodyguard work and anti-kidnapping work. In the global economy, big corporations will do business wherever they can, and it's not much good for company morale if key executives get shot or kidnapped. The bodyguard work in Europe and North America is very specialised; unlike what you might think from watching *The Bodyguard* or *In the Line of Fire*, security in the developed world is mostly preventative. But in Africa, the Middle East, South America and even the Caribbean, visible security is not merely recommended – it is an absolute must.

I was pondering over my future when I got an unexpected phone call from Bob 'Baghdad' Robertson, an ex-British para and a security expert who rang to ask whether I was interested in some work in the Carribbean. Bob had almost been blown up by the IRA in its Warrenpoint ambush in 1983 when a total of eighteen members of the Parachute Regiment died. He had just got out of a helicopter as part of the reinforcements when the second, and most lethal, bomb went off. He'd taken three strides away from the helicopter when he was suddenly blown backwards by the sheer force of the shock wave from the explosion. But Bob bore no grudges against Irish people in general over what happened, and I found him great to work with in Iraq, a real pro and a true gentleman.

Typical of Bob, after the al-Habaniyah ambush he probably thought I could do with some work and needed to get back out into the field. To use the old Western phrase, I needed to get back in the saddle as quickly as possible. He told me about some security contracts that were coming up in the Caribbean and South American that just might suit me. My great friend Mick McCarthy and I applied for a Haiti security contract. It meant providing security for personnel in offices in Port-au-Prince.

Mick and I sent in our resumes and got a phone call the very next day. We were both asked to travel to Dublin for a chat about the contract. Mick was offered a contract almost immediately, but the executive doing the interview felt that I had 'issues with authority'. He bluntly told me he wasn't so sure about involving me with the team in Haiti. So I had to explain what I was about. 'Look, you're right about me having issues, but they're with incompetent leadership, not with authority. I can take orders or instructions,' I explained. 'You don't serve five years in the Foreign Legion without being able to follow the chain of command. But I'm not a politician. If I feel there is a serious problem with something I'll give my honest opinion. That's the way I am.'

This was less than six months after the ambush at al-Habaniyah so I suppose I had a fair bit of street credibility when it came to security issues. After chatting further with the security manager, he told me: 'You're in.' I was assured that my kind of straight talking was exactly what the Haitian mission required. Mick wasn't able to travel for about a month, but I was ready to leave the very next day if necessary. I had several reasons for wanting to get to Haiti straight away. First of all, I wanted to see how I would operate in the field once again – almost to prove to myself that there were no lasting issues from al-Habaniyah. I also wanted to get my mind off the ambush, and sitting at home with little or nothing to do meant that I had been mulling over what had happened in excruciating detail. If I was busy working in Haiti it would take my mind off Iraq. Having rediscovered life as a soldier, which I thought had been lost to me forever back in 1996, I didn't want to lose it a second time.

But, before I travelled to Haiti, I had to go through an obligatory security briefing first. The briefing made the situation sound so bad that we were all braced for the worst – we were half-expecting to see corpses being eaten in the streets by dogs on our way from the airport. It wasn't that bad – but it was bad enough. We were told to treat our mission as Threat Level Five – the only thing ranked above that was the open warfare of Iraq. We would be armed, but all our weapons would have to be licensed by the Haitian authorities. If we were caught with an unlicensed weapon we ran the risk of being arrested and imprisoned. Initially, we would operate as a pretty tightknit team. There would be about eight of us security contractors and we would be protecting between fifteen and eighteen clients.

My ability to take up the job straight away suited the contractor and, just three weeks later, I stepped off the aeroplane into the Caribbean sunshine of Port-au-Prince. We were operating from a hotel just outside the city. There were some very capable operators on the team. There were ex-Legionnaires – myself and Jan, a former Legion paratrooper, and Mick would soon join us. Then there were a couple of other Irish lads and some ex-British Army guys, and we were supported by local Haitian contractors. The team was led at operations level by a former Slovakian Army Colonel, but effective control was with the Irish contractor. I struck up an immediate rapport with a few of the Haitian security operators and being fluent in French helped enormously. I've always believed that there is nothing so priceless as local knowledge. You can have all the security training and military experience in the world, but if you don't know the turf you're at a disadvantage straight away.

There are very few places on earth that can make Iraq look almost normal, but Haiti is certainly one. Never have the contrasts seemed so great between the stomach-churning poverty of a people and the stunning beauty of their country. And believe me, Haiti is beautiful. Wedged on the eastern end of Hispaniola Island which it shares with its neighbour, the Dominican Republic, at first glance Haiti appears to be a tropical paradise to rival Barbados, St Lucia or Tobago. Powder-white beaches, azure waters and, dominating everything, a stunningly-blue Caribbean sky. It truly seems like a little piece of Caribbean heaven. And then you look around you and wonder whether you've landed in Somalia by mistake.

Everywhere you look there is poverty. The roads are flanked by open sewers, there are shanty towns everywhere and the traffic, well, it is so chaotic and congested that for a while I didn't know whether you were supposed to drive on the left or the right. Haiti had that awful, all-pervasive smell of poverty and slumlife that only the Third World generates. And then there is the violence – brutal, savage and almost ritualistically sadistic violence. I can honestly say that Haiti is one of those tragic, terrible countries where human life means close to nothing. It wasn't uncommon to see bodies being left by the roadside, with motorists and pedestrians studiously ignoring what had once been a human being.

There was violence in Iraq, but it was clearly for political reasons. Here, people seemed to kill just for the sake of it. Gangs didn't just shoot their victims – often men would be kept alive for hours while they were slowly beaten to a pulp. It wasn't uncommon for women to be gang-raped until they bled to death. Vicious, gratuitous violence seemed to be a badge of honour for Haitian gangs.

But I'd heard all the stories long before I landed at the airport in Port-au-Prince. When you mention kidnapping, people tend to think about the Middle East or South America. But Haiti's crime gangs had realised that there was serious money to be made from kidnapping and it was now a flourishing business. In December 2005 there had been five hundred kidnappings, most of them involving Haitian nationals, but also European, North American or Middle Eastern business people or members of their families. A portion also included the Haitian staff working for those corporations. So the world's big companies had decided that security was absolutely essential, simply to do business in Haiti.

My immediate impression of Haitians was that they were lovely people, particularly the guys I worked with. They had families to support and were very grateful for the employment and the wages. As a result, most of them took their jobs very seriously – which was just as well because the security situation in Haiti was bad and getting worse. Kidnappings had reached an all-time high the month before I arrived. While we didn't see corpses being eaten in the street, there was enough violence and raw poverty to make Westerners shudder. On one occasion, a female executive in our care was collected from the airport, but she got physically sick at what she saw on her way into Port-au-Prince city and immediately demanded to be driven right back to the airport to catch the first available flight home. There were parts of Port-au-Prince that were truly beautiful, but the city was virtually surrounded by slums and shanty towns where people lived in abject squalor.

Our job essentially focused on keeping the Haitian-based employees safe and looking after visiting executives. I had seen how convoys were run in Iraq and we adopted the same tactics here: fast, aggressive driving to get from point A to point B and not stopping for anything in between, no matter what. Unlike Iraq, where high-powered jeeps and 4x4s stood out like a sore thumb, we were able to use such vehicles in Haiti and they offered massive advantages in height, speed and the ability to get out of a dangerous situa-tion fast. The Haitian police also recognised the vehicles, and they knew we were simply interested in keeping our people safe.

We also applied strict rules about personnel movements. No-one was ever to go anywhere on their own no matter what the excuse. If someone had to go into the city for a busi-ness meeting, they were escorted. It was the same story if someone wanted to head into the city for personal reasons. Personnel were taken to their hotels, where security staff were always in place. Lunch meetings were always escorted and social outings were screened by the security team. The senior executives all had their own private villas and each villa had its own dedicated security team with contact patrols. No-one, no matter what the circumstances, went anywhere without briefing the security teams or accepting protection if it was recommended.

We all carried side-arms which we mostly wore in holsters under our jackets or shirts. This wasn't Baghdad, so there was no walking around on city streets armed like Rambo. Every contractor had a 9mm automatic pistol, mostly the Austrian-made Glock, and teams on convoy duty generally travelled with a pump-action shotgun for heavier fire-power. We also had Uzi submachine guns for extra support. If necessary, we had access to more powerful weaponry, including a Galil (the Israeli-made clone of the Kalash-nikov), and Heckler & Koch MP5 machine pistols. But we were careful not to make our firepower overly visible – better for potential threats to be worried about what you might have than be able to plan for what they knew you were armed with.

For the first four to five months, everything went like clock-work – which was just as well because the violence in Haiti continued to spiral and several times threatened to explode out of control. The UN teams in Haiti began to adopt aggres-sive peace-enforcement tactics, largely in response to the increasing violence being demonstrated by militia groups and gangs. Haitian gangs seemed to demonstrate their will by constantly pushing out the limits of mind-numbing vio-lence. I thought I'd seen the very pit of human cruelty in Iraq, but there were times in Haiti when even I cringed at what was being done. Throughout it all, I often wondered what role drugs and the local voodoo religion played. That and the fact that the poverty in Port-au-Prince's slums was so abject and so sickening that many of the gang leaders probably grew up knowing no other way of life.

Slowly, problems began to escalate for our security details. The company whose employees I protected was rapidly expanding its Haitian commitment, and, slowly but surely, our security resources began to be stretched. The security team expanded to almost twenty personnel. Our practices had to change too: whereas we once took all the staff to a specific restaurant at the same time, executives now wanted to stagger their lunch breaks and go to different restaurants. Personnel also began to undertake journeys in different directions at the same time – for instance we might have one security team heading to the airport while another had to go into the city. But what was most alarming was the way we were suddenly asked to cope with non-priority tasks. On one occasion, we had a security team and their car unavailable for three hours because one VIP was getting her hair done at a city beauty salon precisely at the time we were trying to balance airport runs with lunch escort duties. On another occasion, we were even asked to provide security for a family beach picnic, despite the fact that we hardly had the manpower and security cars to cope with working personnel. Throughout it all, some of the executives treated us more like butlers and chauffeurs than bodyguards. I've seen the corporate ego at close hand and it's not a pretty sight.

Needless to say, there were exceptions, especially at the top level who were perhaps more used to security arrangements, but a handful of junior executives, in particular, were beginning to grate on my nerves. The major difficulty was that client wishes were now beginning to dictate security operations rather than the reverse, which in my opinion was dangerous and unprofessional. Not surprisingly, my Legion training began to show and I started to raise objections. I knew that if a job had to be done, it had to be done right. And in Haiti you simply could not afford to make a single mistake – even to the point of allowing a single security team to feel isolated. My view was that security-threat analysis was paramount. I was now doing most of the security reviews and planning in Port-au-Prince, but there were times when I wondered if anyone was listening to me.

Once, I was asked to do a security analysis of a local carnival in the city where the firm had sponsored a major stand and were paying for some dancers and entertainers as a gesture of goodwill. I spoke to the local police, I checked with our Haitian security contractors and I quickly realised that, in the hot-bed atmosphere of the city, the carnival could easily transform into a security nightmare. But what frightened me most of all was the police warning that barricades were being set up on the outskirts of the city centre and no-one, not even licensed security contractors, would be allowed to carry firearms beyond those lines. So, if we were to provide security for our staff at the carnival, we would be totally unarmed – and massively outnumbered by the crowd.

I recommended that no-one attend the carnival. If the corporate flag had to be flown, I conceded, as a very last resort, that only a handful of staff should go – with a full security team. No-one listened to me. One female executive decided that the workers really ought to see a Haitian carnival at first hand for themselves and that it was vital to 'fly the corporate flag', so to speak. My opinion was that the streets of Port-au-Prince were dangerous enough in daylight – God only knows what was possible with the combination of darkness, heavily armed gangs and street mobs stoked to boiling point with alcohol and drugs. Add a strong dose of voodoo and you have a pretty potent cocktail. Personally, I thought an evening in a hotel restaurant with a steak, a bottle of wine and a heavily armed security guard on both entrance and exit doors was the best way to enjoy Port-au-Prince. But now we had to take a large group of staff executives to witness a carnival that I suspected could at any second erupt into a bloody riot.

My security intelligence briefing strongly advised against going. But the carnival outing was sanctioned and we had no choice but to provide full security. Our entire team went into overdrive to try and protect against the worst-case scenario despite the fact we were massively limited by the fact we would be unarmed. That whole evening we were on edge because we couldn't have stood out more in Port-au-Prince – we looked like rich Yankees, and I'm sure more than a few Haitian gang members watched us that night with dollar signs in their eyes. Luckily, our group was in a stand above the main crowd, which gave us some precious separation from what was happening at street level.

After about an hour at the carnival, which was spent listening to the music and watching the dancers, the whole atmosphere suddenly changed. The shouts and cheers of the crowd began to switch to screams and shouts. And, in the matter of a few minutes, the crowd became a baying mob. People surged forward, crashing against the stands like waves against rocks. The unfortunates on the street, including women and children, began to lose their footing and were forced to the ground where they were crushed and stamped upon.

Our female executive now looked pretty pale beneath her make-up, and every time I saw something horrific occurring in front of us, I turned to look at her. She studiously avoided my gaze and stared off into the distance. But she made no objection when the security detail ordered everyone back to base as quickly as possible. Right in front of us, the crowd was now swaying back and forth, with some people desperately looking for an escape route and others merely intent on violence. The police had only one answer: to wade into the fringes of the crowd wielding clubs and batons in a frenzy. The crowd would shrink back and then, when the police stopped, surge forward once again.

Shortly before we departed, I saw a little boy being crushed. His mother was desperately trying to keep him upright, knowing that if he fell to the ground he would probably be trampled on and killed in a matter of minutes. I moved to the front of the stand and held out my hand, trying to grab the boy and lift him to safety above the mob. But, try as I might, I simply couldn't reach low enough to grab his hand. His mother, likewise, couldn't lift him high enough for me to catch him. The despair in that mother's eyes was heartbreaking. I shouted at one of the other security guards, located below me, to try and reach the little boy but, as the crowd surged again, I lost sight of him. To this day, I don't know whether he lived or died.

I noticed gang members mingling around the fringes of the crowd, careful to avoid the police but watching and noticing every development, like sharks hunting a shoal of fish. There were now also voodoo adherents in the crowd, clearly identifiable by their strange clothing, their ritual chants and the fact that they looked as if they were either in a trance or totally stoned out of their heads. I was hugely relieved when the final signal came for us to pull back and return to base. If a Haitian carnival is someone's idea of a good time, then I think they need serious counselling.

Unfortunately, the carnival was merely the first of numerous incidents which persuaded me that I wouldn't be accepting a second contract here. One leading team member had already left, wanting nothing more to do with the Haitian operation. As the number of civilian clients soared, further security contractors were recruited, but, instead of experienced guys being brought back, we were given lads with no experience or training. They might have been great to provide door security for a nightclub, but this was an entirely different ballgame. Some had army backgrounds, but were still like fish out of water in Haiti.

Most of the new arrivals spoke only English and they began repeatedly to clash with our local Haitian guards. Because I was fluent in French, I found myself being called to sort out problems between our new security contractors and the Haiti lads.

I liked the Haitians, whom I found, for the most part, to be decent and honourable guys. I got very friendly with one of them, Snako, who had the contract to maintain and service our vehicles. Snako was a character – no matter what you needed, he knew someone who could help you. I couldn't believe it when I discovered that he couldn't afford to have his own garage. His work was top-class. No matter time it was, day or night, he was on standby to help us with any problems. I was stunned when Snako asked me to stand as godfather for his newborn child and I accepted because I knew it was a great honour. Yet, despite the fantastic service Snako provided for our operation, he was eventually let go. I could never discover a reason, but it created further doubt in my mind about whether I wanted to stay.

The problems continued to mount. One evening I had just finished my shift and was about to go to bed when I got a call that there was a serious problem at one of the executive villas. I was told it was a major issue so I decided I'd better check it out myself. I drove over and found a newly arrived Irish security contractor standing in the driveway looking very worried. The two Haitian security guards were standing by their hut exchanging vile looks with him. When I queried what was wrong, the Irish guy told me that he was doing a routine patrol when he noticed the two Haitian lads asleep in the hut. It was a serious breach of duties, but it did happen occasionally because the Haitians were often working other jobs on the side to earn extra money. Usually what was done was to give them a warning and stress that, if one needed to sleep, the other had to remain alert.

But the Irish contractor couldn't speak French and judged that the only thing he could do was to shine his torch into the hut to wake up the two sleeping guards. He successfully woke them up, but they didn't appreciate his actions. There was a garbled exchange of views in a mutually incomprehensible mix of English accents and Haitian/French before one of the Haitian guards picked up his shotgun and swung it over his shoulder. In doing so, he brought the barrel to bear on the Irish contractor, effectively trying to intimidate him. When I heard this, I knew I had no choice but to contact the Haitian security liaison officer and get the two guards relieved. But the Irish lad simply hadn't known what to do and I felt annoyed that I'd been forced to come out and sort out someone else's problem.

A short time later, a similar incident occurred, again with a contractor who spoke no French and who had no foreign work experience. He explained that one of the Haitian guards had actually levelled his shotgun at him – and he had replied by levelling his submachine-gun in reply. The stand-off lasted for several minutes before both sides backed down. I was furious that things had gone that far. 'Why the fuck didn't you tell me this earlier?' I asked him 'This is serious shit.' He said that he presumed such stand-offs were part-and-parcel of life in Haiti. I felt I shouldn't have to explain people's jobs to them.

I also felt I shouldn't have to argue with people about trying to keep them safe. If Haiti was deemed to be Threat Level Five, then it should be treated as such. I know some of the executives felt that living under our security restrictions was like being under effective house arrest. But it was like that for the security personnel too. In fact, it was worse. I worked longer shifts in Haiti than I did in

Baghdad – working fourteen or sixteen hours straight was not unusual. We may have been staying in a potential Caribbean paradise but we had to endure regular power cuts, generally had only one decent meal a day and no social life. Our only way of relaxing was to play cards and have a few beers late at night – and one executive even managed to complain about that! So when executives muttered about how tough life was for them, I had to bite my tongue not to remind them that our security personnel were up before them in the morning and stayed on duty throughout the night while they slept in their beds.

The executives seemed totally absorbed by their own concerns – who got to go to a better restaurant than them, who had a longer shopping break and why they couldn't go to the beauty salon when their friend had. For them, it seemed to equate to a promotion competition. For me, Haiti was all about trying to stay alive and get back home safe.

Every time one of our security contractors went out on protection duties they brought a special bag with them we had nicknamed our 'break for the border' kit. This included a passport and sufficient dollars to bribe your way across the border into the Dominican Republic. We knew that, if we ever ran into trouble where we had to use our weapons to protect the personnel in our care, it was essential to get out of Haiti as fast as possible. 'Get the fuck to the border and worry about everything else later,' was the maxim we drilled into every contractor on arrival in Port-au-Prince. Even if we were a hundred percent correct in any contact action we took, the Haitian authorities would, at the very least, throw the entire security team into prison pending a hearing. And survival in a Haitain prison – particularly for a western security contractor – was very much a matter of chance.

The other huge concern I had was that while major attention was being focused on the needs and demands of the corporate executives, we were being treated as an afterthought – in particular, the absolute priority that all of our firearms should be legally held and certified. Without a correct permit, a security contractor, whether Haitian of European, could be arrested or even shot if caught by the police.

I was horrified to discover that several of our weapons had not had their permits renewed, whether through negligence or simply a pathetic effort to try and cut corners and save a few dollars. One day, one of our security patrols was stopped at a Haitian police checkpoint and it emerged that a handgun did not have a permit. It was our nightmare scenario come true. The only thing that averted a potentially nasty showdown was the fact that I knew one of the Haitian policemen – and we had enough dollars on us to bribe our way back to the security compound. I was appalled that while we were trying to protect people, our personal safety was being neglected. To add insult to injury, the company then refused to reimburse us for the bribe money.

I knew I'd had enough. It had been an incredible experience, but it was time for me to head home. I was fortunate in that, because of the al-Habaniyah injuries, I needed long-term treatment, including some plastic surgery. So that gave me the perfect excuse to leave. And, to be blunt, I wasn't sorry to go. I received more gratitude for the protection I offered from a $5 a day Iraqi civilian truck driver than I ever did from some of these well-heeled executives. If these people needed to be persuaded about the importance of their own security, well, that was their look-out. I was a soldier, not a nursemaid.

15

JE NE REGRETTE RIEN

Like most ex-Legionnaires, I don't believe in regrets. As most soldiers who've worn the *Kepi Blanc* will tell you, Edith Piaf's '*Je Ne Regrette Rien*' was not only a great ballad, it seemed to encapsulate perfectly the mentality of all those who served in the Legion. I've lived my life the way I've lived my life. Some things I'm very proud of, and, yes, there are quite a few things that I wish hadn't happened. But I try to learn from them and move on. Usually my life moves in a direction completely unknown to me. I can only laugh when I hear people talk about career paths or the road-map for their lives. I know as much today about where my life is headed as I did twenty years ago.

I haven't a clue where my life would have taken me if I hadn't walked into that Cavan bookshop all those years ago. But I wouldn't swap my experiences in the Foreign Legion for anything – not for gold or cash. I proved more to myself – about myself – in completing the Legion basic training than in anything before or since in my life. I learned about courage, about loyalty, about decency, about the corruption of power and about how, sometimes, life can truly be a game of chance. I learned about how tragedy can bring out the best – and the worst – in people. I learned that bravery is something that can never be predicted or assumed. And I learned that, in warfare, it is the non-combatants who pay the worst bill. The wives, mothers, grandparents and children are those who pay the true, terrible cost of conflict. I've seen too, at first hand, the proof of what the Duke of Wellington admitted when he said that it was just as well war was such a terrible thing so that soldiers wouldn't get too fond of it.

Most of all, the French Foreign Legion taught me about myself: about never quitting and always being true to who you are. It's a lesson I've always tried to keep in mind when other people tell me how I can and cannot live my life.

I read a book by a former Legionnaire last year who wrote that, at the end of the day, almost everyone comes away from the Legion disappointed. It may seem cynical, perhaps even depressing. But I have to agree. I never for one second regretted wearing the Kepi, but having to walk away from the Legion broke my heart and, while I didn't realise it back then, launched me into eight wasted years. In the Legion I became a soldier, a very proud one. Back home I tried to be a businessman and then an ordinary labourer – and I forgot the guiding principle of always being true to yourself.

In general, I don't talk much about the Legion for a number of reasons. Sometimes, in a pub or at a party, when people hear about the Legion and the work that I've done in Iraq or Haiti, they nudge each other and the word 'mercenary' is whispered behind my back, often just loud enough for me to hear. The usual questions then are: 'How many people have you killed?' or 'Are you afraid to die?' For the most part, I try to politely change the subject. Ten years ago I'd probably have given the person a verbal lashing or else gone berserk. Now, I try my best to smile back and yet leave the question unanswered – not because I don't have answers, but I don't think my experiences in Bosnia, Avignon, Cambodia or Iraq should be reduced to the level of pub conversation.

I am not a mercenary and neither am I a gun for hire. I'm a security contractor – a potent contractor, admittedly, but nothing more. I have always considered a mercenary to be someone who will take money for anything – to change lawful governments, attack corporate interests or even fight 'dirty' wars. I will not work for anyone who is trying to undermine or overthrow a legitimate, elected government. Similarly, I refuse any so-called 'dirty' or subversive work, no matter how much money people are prepared to put on the table. I have a pretty strong moral standard and, without exception, I can look back on the work I've undertaken without the merest hint of regret. I offer my services to protect people, whether it is a humble American sanitary engineer trying to repair a sewer in Basra or a Jordanian father-of-five who is working as a cook for civilian contractors in Baghdad. I am a security contractor and my job has always been to protect people, not to hurt them. I like to think of it as someone standing between the sheep and the wolves. But I make no apologies about using my rifle against those who would maim, execute or brutalise the innocent. I have no regrets about what I've done to protect the people in my care – if you take a rifle in your hands to hurt the people I'm protecting, you've already accepted the consequences of that action.

My life is the sum of my choices, some good and some bad. I understand that people are fascinated by the places I've been and, most of all, by the Foreign Legion. But all too often if you try and explain to people what it was really like, they only want to hear about the shootings and the lifestyle. Quite often, if it's mentioned in a pub that you were in the Legion, people assume you must be some kind of hard man and there's always one idiot who has consumed about ten pints and wants nothing better than to prove, in front of his girlfriend and mates, of course, that he's tougher that you.

The truth is that I've had to be tough to take some of the things thrown at me in the Legion and Iraq. But I'm not a hard man. I learned in the Legion that this type of Rambo-character inevitably ends up in trouble. I've tried my best to be a low-profile operator. Yet it's a demanding field we operate in and often there is little time for sentiment, even when people you care about end up badly injured or even killed.

In Iraq, I lost many friends – Nick Pears, Saito, Yves, Sean, Arkan and Wisam. As I write this book, I've heard that my Norwegian buddy, Gere, was badly burned in an ambush where the insurgents used the new type of copper-fuel cell IED. I don't know if he'll ever be able to use his hands properly again. But the fact remains that, the next day, there are still people to be protected, there are still convoys of fuel, medicines, foodstuffs, ammunition and machinery to be delivered. If Iraq is to avoid the bloodbath of a civil war which now looms, those convoys have to keep getting through.

But the price is very high and the butcher's bill paid by security contractors is never highlighted and never mentioned. Analysts can reel off statistics of the number of US or British soldiers killed or injured in Iraq, but do they know how many private contractors have died? What surprises me most is that, from my information, almost as many private contractors have been killed in Iraq as Coalition soldiers. And most of those find their final resting place in an Iraqi desert or ditch. They don't even get the respect of being flown home in a flag-draped coffin for a funeral. It's very hard to accept that fate for friends and comrades.

How do you explain to someone what it's like to go to the South African embassy to deliver a sworn affidavit to some nameless bureaucrat that you saw one of your friends being killed right in front of you? That you know for certain that Sean L. is definitely dead? And to realise that that affidavit is crucial if his family are ever to be able to claim life insurance payments? This wasn't a statistic that I saw being shot, it was my friend Sean – and when he was killed I thought we were all going to die alongside him. How do you react when some embassy bureaucrat, sitting behind a desk, who has never handled a rifle or trusted another person with their life, asks how you know *for certain* that your friend is dead? All for a lousy insurance payment!

Yes, I have shot people. And, yes, I have thought about it afterwards. But what has helped me to square it with myself has been the fact that, in every single case, if I hadn't fired those shots the person would have killed me, one of my mates or one of the people I was charged to protect. It's as simple as that. I have never deliberately gone out of my way to harm anyone. If the convoys I protect aren't fired upon, then I keep my weapon lowered. If I can avoid a confrontation, believe me that is the route I choose to take. As a security contractor, I will take boredom over confrontation every day of the week. But I won't ignore the duty of care I have to the people I protect. If you try to kill them, well, you'll have to kill me first.

Yet there are hundreds, probably thousands, of people determined to judge security contractors without knowing the full facts or having ever attempted to walk a mile in our shoes. To some, we're just mercenaries, the so-called 'dogs of war' who don't even deserve an epitaph – ruthless gunmen who will do anything for money. But try telling that to the doctors, the engineers, the teachers and the ordinary businessmen whom security contractors keep alive on a daily basis in Iraq and a host of other countries worldwide. Even the media, often among the harshest critics of PMCs in Iraq, only operate in parts of Iraq thanks to the work of private security contractors.

From my experience of Iraq, all the people there want is to be able to live their lives in safety, raise their children and work to make a better country. The vast majority, I believe, don't support the insurgents. But there is deep hatred between Sunnis and Shi'ites and that is something the entire world should fear because of its possible consequences. But Iraq is not going to rejoin the world of nations without engineers, doctors, teachers and businessmen, so keeping them safe while they rebuild the country is probably one of the things I'm most proud of. The people of Iraq deserve a better future – God knows, they have suffered more than their share of heartache and tragedy over the past fifty years.

What I don't understand is how their future can be better if the country is abandoned to civil war and sectarian violence. I am not an apologist for the invasion of Iraq – personally, I think the reasons for invasion are highly dubious and I believe that the whole affair could certainly have been handled vastly better. There is no doubt, in my mind, that those who over-saw the invasion and the subsequent transition situation bear a heavy responsibility for the carnage that has followed. But we are now faced with the problem of the aftermath and how best to deal with it. Having worked, lived and almost been killed in Iraq I can say one thing with utter conviction: if the lifeline of convoys is cut, the bloodshed to date will be dwarfed by the resultant slaughter and it could take Iraq an entire generation to recover from the carnage.

As for the moral justification for protection duties in the modern world, well, explain to the family of a businessman kidnapped and brutalised in a place like Haiti that they're the risks you are supposed to run for doing business in the Third World. How can anyone argue that, on moral grounds, you really shouldn't employ armed security con-tractors? I once heard a soldier comment that there is the world we would like to live in and then there's the world we actually reside in. I am part of the latter reality. Like most soldiers, I've experienced the inherent contradictions of military life – loving the comradeship and the lifestyle while hating war and the loss of friends.

No-one knows better than I do that fighting a war is a truly terrible thing – but sometimes not to fight is even worse as the horrors of the Nazi concentration camps fully underlines. Yet, not all wars fall into such easily justified categories; they usually rank as shades of grey rather than simple black-and-white equations. What was Vietnam all about? Or Cambodia? Or the Balkans? Or even World War I?

I have no idea how future generations will classify the US-led invasion of Iraq. I have my own suspicions but, for my part, I am proud to have served alongside men like Yves, Sean, Arkan and Wisam. They died trying to protect the innocent and, throughout history, I believe that has always been something very noble.

After Iraq I promised myself a decent break to recuperate and recharge my batteries. Yet, a few months later, I ended up in Haiti. Now, I'm trying to enjoy that long-overdue rest.

There are times when I feel that my face and body are like a roadmap of the most dangerous places in the world – a wound on my cheek reminding me of Iraq and a scar on my side reminding me of Bosnia. I've only been three years working in the private security sector, but it feels more like ten. I also know that I've paid a pretty high price for the life I've led, and I know my family have as well. One of my biggest regrets is the worry that my parents and sisters have gone through, particularly over what happened in Iraq.

I suppose the biggest price I've paid has been in terms of relationships. A lot of the guys I went to school with are now married, have several children and are well into their mortgages. But I've never been in one place long enough, or been happy enough there, to start any meaningful relationship. Any female contacts have been casual or extremely short-term. It's no coincidence that one of the most meaningful relationships I've ever had has been while I was in Ireland over the past year, resting and recuperating. To be honest, it's something I have missed, and, at thirty-seven, you realise that you're getting a little old to be still living the same kind of single life you enjoyed in your late teens and twenties. But the work I do is far easier to handle by being single. I have my doubts about expecting a wife to be at home trying to live a normal life while knowing their partner is in danger in the Middle East or Africa. It is not an easy life to live, yet tens of thousands of women (and a few men) do it worldwide.

But both the Legion and working as a security contractor have instilled a wanderlust in me and I know that there's no changing that now. I think it will probably take me overseas again in the near future. I don't know precisely where, but I do know that, perhaps sadly, there is no shortage of such work, particularly from corporate clients. I've already had offers from companies in Africa, the Caribbean, the Middle East and even Asia. To be honest, I don't know precisely where I'll eventually end up. I suppose part of the tragedy of the modern world is that decent people doing valuable work still need men with guns to protect them. And while that threat continues to exist, I will never apologise about loading my rifle and standing between the sheep and the wolves.

APPENDIX

FRENCH FOREIGN LEGION

Founded in 1831 by France's 'Citizen King' Louis-Philippe, the French Foreign Legion – or La Légion Étrangère – has established itself as one of the world's most famous elite fighting forces. From Algeria to Mexico and from the battlefields of World War II to Vietnam, the Foreign Legion earned a reputation for toughness, courage and skill that has made it a household name worldwide. The unique character of the Legion has also been the focus of numerous books and films for decades, ranging from *Beau Geste* to *Devil's Guard* and *March or Die*.

The Legion is perhaps most famous for the astonishing courage its soldiers have demonstrated in defeats rather than for its major military victories. Like the Spartans at Thermopylae, defiance in the face of overwhelming odds became the Legion mantra. Over its 176-year history the Legion provided the French with a tough, professional corps within its largely conscript army. Ironically, the future and ethos of the Legion came under serious threat only when the French Army, since 1995, began to move increasingly towards a professional, volunteer model.

But, to date, the corps, created in the wake of France's July Revolution of 1830, has survived one kingdom, one empire and three republics, not to mention two world wars and the painful dismantling of France's colonial empire, which it was originally created to protect. The Legion even managed to survive a revolt by one of its elite units against the French Government.

Over its first decades, the Foreign Legion was primarily entrusted with protecting France's overseas colonial interests, from Africa to Asia and South America. It also served a useful role in helping to remove 'anti-social elements' from both French and European society – a fact King Louis-Philippe was fully cognisant of in the wake of Parisian riots. Yet, for the major part of its existence, the Legion could not legally be used on French soil, a fact that first caused major problems during the Franco–Prussian War in 1870 when the 5th Foreign Battalion had to be formed to allow the Legion be used to defend France against the Prussians. In one major engagement, at Orleans on 10 October 1870, the battalion was virtually wiped out. Interestingly, one of the survivors was Sous-Lieutenant Kara – one Prince Piotr Karageorgevich, who went on to become King Peter I of Serbia.

The Legion's almost cult status began with its deployment to Algeria and Mexico in the 1840s. Algeria would prove to be the Legion's home for more than a century and its base at Sidi-Bel-Abbès became the stuff of legend. From the outset, the Legion was sent to do the work the French Army resented or resisted, a memory that lingers to this day. In one interesting throw-back to the French Revolution, the Legion had a special green-clad uniform for recruits to its 1st Regiment Étrangère (RE). This was because, like the famous guards of the Bastille, the 1st RE was entirely manned by Swiss recruits. But, by 1859, the lack of sufficient Swiss soldiers forced the 1st RE to be incorporated into the remainder of the Legion and the green uniforms to be abandoned. From then on, the Legion's colours would be blue and red.

It was in Mexico that the Legion found its fame when a small force of just sixty-five men led by thirty-five-year-old Capitaine Jean Danjou fought heroically against a huge Mexican force of more than a thousand troops on 30 April 1863. The Legion unit fought with reckless bravery and only three Legionnaires were captured alive. Danjou's courage made the Battle of Camarón legendary – the Capitaine's wooden artificial hand was later recovered from the Mexicans and became one of the Legion's most iconic treasures. To this day, the Legion marks 30 April as Camarón Day. The Legion's legend was also established through various colonial conflicts in Tunisia, Algeria and Indochina before World War I. The Legion even fought with distinction in the Crimea in the 1850s. But the Legion's longest and bloodiest campaign was against rebellious tribes in North Africa, as portrayed in *Beau Geste*. But, just as the Legion was on the verge of finally quelling unrest in North Africa, it was summoned back to Europe for the Great War. The Legion fought alongside regular French troops in the trenches and even played a key role in the Battle of Verdun, one of the conflict's bloodiest campaigns.

In World War II, the Legion saw itself split, with part remaining loyal to the Vichy Government and a larger portion supporting General Charles de Gaulle and the Free French Forces. In North Africa, Legion units fought heroically against German Commander Erwin Rommel's Afrika Corps at Bir Hakeim in an epic battle, then went on to fight in Italy and during the liberation of France.

After World War II, the Legion was at the receiving end of France's failed attempt to restore the pre-war status quo to its empire. In Vietnam – once part of French Indochina – the Legion fought an increasingly bitter war with the Chinese-backed Vietminh. Despite some successes under its Commander, General Jean de Lattre de Tassigny, the Legion lacked the strength to tackle the Vietminh and was receiving mixed political signals from Paris about the overall aims of the conflict. In 1954, under General Henri Navarre, the Legion would suffer a catastrophic defeat at Dien Bien Phu, an isolated valley between Tonkin and Laos. But Legion lore was again enhanced when it emerged that Legionnaires had volunteered to be parachuted into the beleaguered garrison despite having no hope of evacuation or final victory. Legion paras, under Colonels Pierre Langlais and Marcel Bigeard, fought with Spartan courage – one famous story tells of German-born Legionnaires singing a slow marching cadence as they went into battle against hopeless odds.

Ironically, it was France's colonial collapse that posed the greatest threat to the future of the Legion. When President de Gaulle ordered French forces out of Algeria in 1961 following a vicious guerrilla war, sections of the Legion revolted. Indochina may have been a colony, but Algeria was considered by many to be part of France, as much as Brittany or the Loire were. From a Legion perspective, the abandonment of Algeria meant the Legion had fought a savage guerrilla war for more than five years for nothing. The decision also meant the Legion had to abandon Sidi-Bel-Abbès, which had been its famous home for more than a century. Elements of the Legion revolted and France was greeted with the spectre of regular Army units confronting Legion paratroopers. But the revolt was put down and the Legion's 1st Regiment Étrangère Parachutiste (REP), which had led the insurrection, was formally disbanded. In 1962 the Legion was reduced in overall numbers as France axed its so-called 'Army of Africa'. But the Legion avoided the fate of other famous French colonial units such as the Zouaves and Spahis which were totally abolished.

Sidi-Bel-Abbès was indeed abandoned but its famous monument to the Legion dead, a giant bronze sphere held aloft by four sculpted Legionnaires, was dismantled and re-erected at the Legion's new home at Aubagne in southern Provence. The Legion even re-created its famous 'Sacred Way', a special path used by military colour parades, in front of it most cherished memorial and the sculpture itself was deliberately turned to face the south, towards Algeria, where so many thousands of Legionnaires lay buried in the Saharan sand.

By the 1970s and 1980s, the Legion was once again proving its mettle with various African deployments and an increasing amount of anti-terrorism work. In 1978, it was the Legion that was assigned to Kolwezi in the Congo where rebels had captured European hostages and mining interests. In 1990, the Legion was sent to the Middle East to take part in Operation Desert Storm, and the Allied Commander, General Norman Schwarzkopf, hailed the Legion as one of the toughest units he had ever worked with. Subsequent assignments included detachments being sent to Cambodia, the Balkans following the Yugoslav civil war, and war-torn parts of Africa.

Ten percent of Legion officers are promoted through the ranks while less than one-quarter of ordinary Legionnaires are French-born. Following World War II, the Legion boasted a hard-core of former German soldiers and currently it has a large proportion of troops with military experience from the former Soviet Bloc. Legionnaires are allowed to apply for French citizenship after three years' service. Terms of service with the Legion are five years in duration, reduced from the previous norm of seven years.

Since 1831, the Legion has fought in almost thirty different countries and lost more than 900 officers, 3300 sous-officers and 31,500 Legionnaires.

SECURITY CONTRACTORS AND MERCENARIES

The global security industry, ranging from the domestic security sector through to the private military industry, is estimated to be worth in excess of one trillion dollars, rivalling even the total cost of global healthcare. And, in a notoriously difficult area to define, the broad agreement is that personnel involved in protection duties for convoys, hospitals or key VIPs are *security contractors*, while those who take an active or non-defensive role in a conflict are regarded as *mercenaries*. However, legislation and international agreements often fail to distinguish between them.

Under the strict definition of the Geneva Convention, any American civilian who worked as an armed security contractor could not be regarded as a mercenary while the US was in total or partial control of Iraq. But when the US authorities handed over legal responsibility for the country to the new Iraqi Government the situation changed. US soldiers are not affected because they are regarded as being 'a party to the conflict', a status which has been endorsed by the Iraqi Government. But it has been argued that unless civilian contractors declare themselves to be residents of Iraq – and thus become 'a resident of a party to the conflict' – they run the risk of being technically regarded as mercenaries. If a person is a resident of Iraq, even on a temporary basis, they can claim status under Article 47-D of the Geneva Convention. Under the Geneva Convention (1949), a mercenary is not accorded the rights of a combatant or prisoner of war. The convention deems that a mercenary is a soldier who fights for private gain and has no direct political or ideological connection to the conflict. However, the legislation that governs this area – Protocol GC1 (1977) – has not been signed by numerous countries, including the United States.

The status of 'mercenary' can only be determined by a domestic court which must place an individual on trial and offer them the full recourse of the law. If, after a fair trial, the person is ruled to be a mercenary then they formally lose all the protection offered to combatants under the Geneva Convention. Yet the issue of defining a mercenary remains hugely controversial, with some critics even warning that the Foreign Legion and the British Army's famous Gurkha Regiment, which boasts recruits from

Nepal, could fall foul of the new interpretations.

Despite the United Nations adopting a special Mercenary Convention in 1989, it has been argued by neutral and non-aligned countries that the legislation has totally ignored the growing use of Private Military Companies (PMCs) by sovereign states. Because these firms act on behalf of legitimate, elected Governments this situation has thrown many of the interpretations under the Geneva Convention into uncertainty; a case in point is the personal bodyguard of the Afghan President, Hamid Kharzai, a repeated target of Taliban assassination attempts, who is provided from the ranks of PMCs.

The confusion over PMCs has been so great that the South African Government, which passed anti-mercenary legislation in 1998, is now having formally to review the laws in the light of what has been happening in Iraq and Afghanistan, and the fact that its citizens were employed on protection duties there.

The issue is extremely wide-ranging. Switzerland only permits its citizens to legally serve in one overseas military force – the Pope's Vatican Guard. Austria, many of whose citizens fought with the Foreign Legion after World War II, now enforces legislation which strips a citizen of their Austrian citizenship if they serve in any overseas army.

The Iraq War is the first major conflict in which PMCs have come to widespread public attention, though they have been involved in conflicts since the 1970s, mostly in Africa. The best-known PMCs are Executive Outcomes, which hit the headlines over its work in Angola and Sierra Leone, as well as Sandline International, which was involved in controversial contracts in New Guinea and Sierra Leone.

The advantages of PMCs to Governments are significant. A country can effectively minimise the risks to its own military personnel, distance itself from a specific conflict, reduce the cost of involvement in any warzone and, in certain cases, avoid adverse publicity while ensuring a specific political goal is achieved.

In Iraq, the first PMC to make international headlines was Blackwater-USA which had four security contractors killed outside Fallujah in March 2004 while they were protecting a food convoy. The vicious desecration of the bodies by insurgents – one of which was burned, dismembered and then hung from a bridge – is believed to have been a key factor in the Coalition's decision to try and stabilise the Fallujah pocket. Weeks of savage fighting followed.

Ironically, despite the growing controversy over the work of PMCs in conflict zones, the UN has itself worked with PMCs, for example when it awarded a logistical contract to Executive Outcomes in sub-Saharan Africa. A key British Foreign Office consultative report admitted that, in the modern world, it was cheaper and often more effective for a Government to hire a PMC than assign its own troops to an overseas project. However, former UN Secretary-General, Kofi Annan, ruled against using PMCs to support future UN missions and opted to retain the time-tested formula of using mandates to secure troops from UN member states.

CAMBODIA

Like its neighbour, Vietnam, Cambodia paid an horrific price for European colonial encroachment and the loss of centralised control. Despite being home to the powerful Angkor kingdom in the ninth century, which allowed the native Khmer people to develop a highly-regulated and centralised state, Cambodia came under French influence from the early nineteenth century. By 1863, Cambodia had become a formal part of France's Indochinese colonial empire, which was run from Vietnam.

The destruction of local European power by the Japanese during

World War II fostered an independence movement and Cambodia was declared a sovereign state and monarchy in 1953. But King Sihanouk failed to maintain broad-based support and the savage war in neighbouring Vietnam led to an increasing polarisation between left-wing and right-wing parties. In 1970 King Sihanouk was overthrown by his own army and fled to exile in China. Just one year earlier, the US had commenced massive bombing of suspected Communist bases and supply routes to Vietnam inside the Cambodian border. Incursions from Vietnam into Cambodia by US forces and their Army of Republic of Vietnam (ARVN) allies proved bloody failures.

Cambodia's Communist-orientated rebels, the Khmer Rouge, were massively boosted by that campaign and took over the entire country in 1975. The Cambodian capital, Phnom Penh, fell to the Khmer Rouge in April 1975, just a fortnight before Saigon fell to the Vietcong/Vietminh. The Khmer leader, Pol Pot, was determined to transform Cambodia into a Maoist, agrarian state. His four-year campaign to return Cambodia effectively to the Stone Age ranks as one of the most brutal experiments ever attempted by humankind. It is estimated that more than one million people died between 1975 and 1979 from executions, torture, famine and disease. The ability to read or write often earned a death sentence.

Ironically, it was the post 1975 Vietnam state, a unified Communist state incorporating north and south, who decided to end the experiment and, tired of Khmer Rouge attacks along their border, mounted a massive invasion. The Khmer forces fled to the mountains and Cambodia then began a vicious twelve-year civil war as the Khmer Rouge, with direct support from China and Thailand and indirect US aid (for those anti-Vietnamese forces now subsumed into the Khmer Rouge), fought against the Vietnamese-supported Government in Phnom

Penh. By 1991, contacts about peace talks had started and, in mid-1993, elections were staged under United Nations supervision. Military support for the UN was provided by French Foreign Legion detachments among others. King Sihanouk returned to power and a new Constitution was drawn up. While initially outside the power process, the Khmer Rouge were gradually enticed away from armed insurrection. Pol Pot died while under house arrest at a Khmer Rouge military camp before he could be brought to trial for crimes against humanity.

BOSNIA

The Yugoslav Federation, so carefully crafted by Josip Broz Tito, collapsed in 1990/1991 amid growing nationalism from both the Serbs and Croats. But the major battleground soon proved to be Bosnia, the only European country with a traditional Muslim majority. Arms embargoes on Yugoslavia only served to favour the Serbs, who had secured the bulk of the former Yugoslav Army's equipment, and the Croats, who had been able to arm before blockades were enforced. As both the Croats and Serbs scrambled for territorial gains, the conflict quickly adopted a vicious ethnic tone, with civilians from rival ethnic groups being forced out of territories where they were in a minority. The phrase 'ethnic cleansing' re-appeared in the European vocabulary with chilling consequences.

The conflict also saw shadowy criminal groups transform themselves into special paramilitary units, often marked by their brutality and deliberate targeting of civilians. One Serb group, Arkan's Tigers, even posed for publicity photos while they ethnically cleansed certain areas of Muslims.

In July 1995, Bosnian Serb forces, under General Ratko Mladic, occupied the UN 'safe haven' of Srebrenica in eastern Bosnia and, after persuading a Dutch UN detachment not to engage in armed resistance, deported 20,000 Bosnians, mostly Muslims. More than 8,000 Bosnian men and boys were subsequently executed in the worst mass-killing in Europe since World War II. Thousands of Bosnian women were subjected to mass rapes and sexual assaults. The lightly-armed Dutch Army unit – DutchBatt – didn't fire a shot to prevent the Serb deportations.

The horrific treatment of Bosnian Muslims sparked outrage throughout the Islamic world and thousands of volunteers flocked to the Balkans to serve in the 7th Bosnian Muslim Brigade in the Zenica area. Fighting raged in Bosnia between March 1992 and November 1995 when NATO, under Operation Deliberate Force, finally used heavy air-strikes to drive back Serb and Bosnian Serb forces.

The peace agreement signed at Dayton, Ohio, in the US finally ended the conflict in December 1995 and established the modern borders of the former Yugoslav states. War crimes tribunals are still underway in The Hague over the atrocities in Bosnia – though both former Bosnian Serb leaders, Radovan Karadic and Ratko Mladic, the two most wanted men, remain at large. A $10 million reward is on offer from the US Government for the arrest of both of them, but they are both still living in Serbia. Recent estimates have indicated that the war cost 100,000 lives – mostly civilian – and left 1.8 million people displaced in the worst European conflict since World War II.

IRAQ

Few nations on earth have witnessed the bloodshed Iraq has encountered over the past two thousand years. The Hittite, Assyrian, Babylonian, Persian, Greek, Roman, Byzantine and then successive Islamic empires all fought fierce battles for control of the vital Tigris and Euphrates valleys.

In early 2003, US-led Coalition forces poured into Iraq claiming that President Saddam Hussein had failed to comply with United Nations directions about weapons of mass destruction (WMD). The invasion appeared, to many, to be unfinished business that the US authorities had with Saddam dating back to his invasion of Kuwait in 1990 and the subsequent Gulf War in 1991. President George H. Bush ordered Coalition forces to drive Iraqi troops out of Kuwait in 1991, but failed to topple Saddam's regime when he halted Coalition troops just inside the Iraqi border. This became a factor in the US presidential election, which George H. Bush lost in 1992 to Bill Clinton. However, his son, George W. Bush, was US President twelve years later when a fresh Coalition assault was mounted on Iraq, starting on 20 March 2003. After initial setbacks, the Coalition forces pulverised what organised Iraqi resistance was mounted, and Saddam's regime collapsed from within. Saddam went on the run. The so-called 'shock and awe' military doctrine of US Vice-President Dick Cheney and US Defence Secretary Donald Rumsfeld appeared justified.

The 'shock and awe' doctrine has been described my many as the modern equivalent of Blitzkrieg. It uses maximum firepower, mobility, technology and force integration (Army, Navy and Air Force) to force opponents off-balance and then to systematically destroy their ability to fight back. However, its major advantage is also its major weakness – it relies on far less manpower reserves than traditional offensive systems. If 'shock and awe' doesn't completely end a conflict, an army designed and trained for mobility and flexibility is suddenly forced to fight an entirely different kind of irregular war, often one for which it is totally unsuited.

A few months later, Saddam's two sons, Uday and Qusay, his heirs apparent, were both killed having refused to surrender to US forces. Saddam himself was captured on 13 December 2003, hiding in a hole in the ground near a farmhouse, and was later put on trial for war crimes. He was convicted and sentenced to death by an Iraqi Court.

The Iraqi Government later refused calls for the execution to be delayed, pending appeals. Saddam's hanging on 30 December 2006 caused outrage after video-footage was released on the Internet which showed the dictator being taunted by Shi'ite guards as he stood on the gallows just seconds from death. Secret footage was subsequently released of the dead dictator's body lying on a hospital gurney after the execution.

Long before the execution, hardline members of the former Revolutionary Guard, as well as Ba'ath Party members and an increasing influx of fundamentalist Islamic fighters from overseas began to increase pressure on the already-stretched Coalition forces. A growing number of suicide-bomb attacks was followed by increasingly sophisticated ambushes using Improved Explosive Devices (IEDs). These shaped charges were lethal to everything but the most heavily armoured vehicle, and the lightly armed Coalition troops, many driving Humvees and Land Rovers, suddenly became very vulnerable. The Coalition's reliance on the high-tech weaponry at the centre of the 'shock and awe' doctrine was now of little help in dealing with the insurgents and Coalition manpower became a major issue with US forces struggling to balance commitments to Iraq with those in Afghanistan, the Middle East and Korea. The bipartisan Iraq Study Group in the US, led by former US Secretary of State James Baker, was highly critical of the policies adopted and warned that a fresh approach had to be developed in Iraq.

But the attacks on Coalition forces went hand-in-hand with attempts by insurgents to fan hatreds between Iraq's Sunni and Shi'ite communities. The conflict became so bitter that mosques were attacked and at hospitals injured patients abducted and killed if they were mistakenly taken to a hospital controlled by the rival sect. The sectarian violence led to thousands of Iraqis fleeing their homes for the safety of enclaves controlled by their own religious grouping, either Shia or Sunni. By October 2006, the US was losing troops at a rate not witnessed since Vietnam, with 102 killed that month alone. British casualties in southern Iraq had also begun to rise. The Bush Government, following a virtual collapse in Republican support in the November 2006 US mid-term elections which cost the party control of both the US Senate and Congress, then signalled its determination to pursue new strategies to ease tensions in Iraq, insisting that 20,000 extra US troops could help secure Baghdad. Both Iran and Syria are expected to be consulted about future policies as Iraq teeters on the brink of all-out civil war.

HAITI

The Carribbean state of Haiti is the only country in the world to have staged a successful slave revolt. But the country, which is part of the island of Hispaniola, the other part being the Dominican Republic, ranks as one of the poorest and most violent in the world. Ironically, when Christopher Columbus thought he had landed in the Far East – and was later credited with 'finding' North America – he had actually landed on Hispaniola.

Today, Haiti's 8.5 million inhabitants suffer from one of the world's highest infant mortality rates and a life expectancy that is marginally lower than that of war-torn Somalia. The United Nations ranks Haiti as 153rd of 177 countries in its Human Development Index (HDI). Horrifically, it is now estimated that more than 80 percent of all Haitians live in abject poverty.

While Haiti is predominantly a Roman Catholic country, it is also famed for the popularity of voodoo amongst its citizens. Haitian criminal gangs have also earned a reputation for almost sadistic violence, particularly in turf wars or drug racketing.

The country's recent history has been marked by brutal oppression under dictators Francois 'Papa Doc' Duvalier and his son, Jean-Claude 'Baby Doc' Duvalier, military coups, popular uprisings and two US-led invasions. Elections in 2005 were marked by savage fighting between rival factions – with kidnapping, particularly of Western officials, emerging as one of the most lucrative activities. Haiti is now ranked as one of the most dangerous places in the world for travellers.

Ironically, despite the dangers, Haiti remains a major focus for some of the world's biggest companies, ranging from telecommunications firms to energy combines and food retailers. The growing interest of energy firms in Haiti has been underpinned by Venezuela's strong support of the country over recent years, including generous oil-supply contracts. Many companies operate in Haiti as part of a commitment to offering Caribbean-wide services – and also to maintain a presence in Haiti so that, should stability emerge and the economy finally revive, they can prevent rivals from staking a claim. Haiti was once regarded as being amongst the Caribbean countries with greatest tourism potential because of its natural beauty and exotic culture. But decades of savage violence and rebellions have virtually wiped out any tourism trade to the point where some cruise companies who use Haiti's tiny uninhabited offshore islands for brief swimming breaks often decline to inform their passengers that they are actually on Haitian soil.

THE KALASHNIKOV

Designed in 1947 by Russian infantryman and weapons specialist, Mikhail Kalashnikov, the AK-47 assault rifle has emerged as arguably the most lethal firearm ever produced. It is estimated that in excess of thirty million AK-47s and its derivatives have been manufactured worldwide. Production is still underway, with variants made in

Russia, Hungary, East Germany, Romania and China. Only the Belgian FN and the US-made M16/Armalite rival the Kalashnikov in terms of such vast production numbers. Even the Israeli Army, after an exhaustive evaluation of weapons worldwide, chose to adopt a clone of the AK-47, the Galil.

The Russian Federal Army's current weapon – the AK-74M – is little more than a modernised AK-47 which fires a smaller calibre 5.45mm-higher velocity round. However, many countries prefer the older rifle with the stopping-power of its heavier 7.62mm round. Worldwide, more than sixty armies use the AK-47, and analysts believe that, with the exception of mankind's adoption of the sword, no other weapon has been used in as many conflicts as the Kalashnikov. The rifle is so rugged and reliable that it will fire from the sub-zero temperatures of Northern Finland through to the sand and dust of Afghanistan and Iraq. The rifle even inspired part of the plotline in a major Hollywood film, *Lord of War*.

Bibliography

Adams, James, *Secret Armies*, London (Atlantic Press, 1988)

Boyd, Douglas, *The French Foreign Legion*, London (Sutton, 2006)

Geddes, John, *Highway to Hell*, London (Century, 2006)

Harvey, Dan, *Peacekeepers*, Dublin (Mentor Books, 2000)

Jenning, Christian, *A Mouthful of Rocks*, London (Bloomsbury, 1990)

McGorman, Evan, *Life in the French Foreign Legion*, London (Hale, 2002)

Murray, Simon, *Legionnaire*, London/Paris (Presidio, 1980/2005)

Penta, Karl, *Have Gun Will Travel*, London (Blake, 2003)

Sloan, Tony, *The Naked Soldier*, London (Vision Books, 2005)

Spicer, Tim, *An Unorthodox Soldier*, London (Mainstream, 2000)

Weverka, Robert, *March or Die*, New York (Mass Market, 1977)

Windrow, Martin, *The Last Valley*, London (Cassell, 2005)

Young, John Robert, *The French Foreign Legion*, London (Thames & Hudson, 1984)

List of Sources

Newspapers: *The Irish Independent*, *The Sunday Independent*, *The Evening Herald*, *The Times* (London), *The Daily Telegraph*, *The Independent* (London), *The Guardian*, *The New York Times*, *The Chicago Sun-Times*, *The New York Post*, *The Pittsburgh Post-Gazette* and *The Jerusalem Post*.
Broadcast: RTE, TV3, BBC, ITV, CNN, NBC, Sky News, County Sound, 96 FM, NewsTalk.